CAMBRIDGE LIBRARY COLLECTION

Books of enduring scholarly value

Physical Sciences

From ancient times, humans have tried to understand the workings of the world around them. The roots of modern physical science go back to the very earliest mechanical devices such as levers and rollers, the mixing of paints and dyes, and the importance of the heavenly bodies in early religious observance and navigation. The physical sciences as we know them today began to emerge as independent academic subjects during the early modern period, in the work of Newton and other 'natural philosophers', and numerous sub-disciplines developed during the centuries that followed. This part of the Cambridge Library Collection is devoted to landmark publications in this area which will be of interest to historians of science concerned with individual scientists, particular discoveries, and advances in scientific method, or with the establishment and development of scientific institutions around the world.

Conversations on Chemistry

Jane Haldimand Marcet (1769–1858) was a pioneer in the field of education who wrote accessible introductory books on science and economics. Noting that women's education 'is seldom calculated to prepare their minds for abstract ideas', she resolved to write books that would inform, entertain and improve a generation of female readers. First published anonymously in 1805, her two-volume *Conversations on Chemistry* swiftly became a standard primer, going through sixteen editions in England alone, and was cited by Michael Faraday as having greatly influenced him. Presented as a series of discussions between a fictional tutor, Mrs. Bryan, and her two female students, the flighty Caroline and earnest Emily, *Conversations* combines entertaining banter with a clear and concise explanation of scientific theories of the day. Volume 2, on 'Compound Bodies', contains spirited exchanges on topics including 'shells and chalk', borax, decomposing vegetables and 'animal economy' which will interest historians of both science and education. For more information on this author, see http://orlando.cambridge.org/public/svPeople?person_id=marcja

Cambridge University Press has long been a pioneer in the reissuing of out-of-print titles from its own backlist, producing digital reprints of books that are still sought after by scholars and students but could not be reprinted economically using traditional technology. The Cambridge Library Collection extends this activity to a wider range of books which are still of importance to researchers and professionals, either for the source material they contain, or as landmarks in the history of their academic discipline.

Drawing from the world-renowned collections in the Cambridge University Library, and guided by the advice of experts in each subject area, Cambridge University Press is using state-of-the-art scanning machines in its own Printing House to capture the content of each book selected for inclusion. The files are processed to give a consistently clear, crisp image, and the books finished to the high quality standard for which the Press is recognised around the world. The latest print-on-demand technology ensures that the books will remain available indefinitely, and that orders for single or multiple copies can quickly be supplied.

The Cambridge Library Collection will bring back to life books of enduring scholarly value (including out-of-copyright works originally issued by other publishers) across a wide range of disciplines in the humanities and social sciences and in science and technology.

Conversations on Chemistry

In which the Elements of that Science are Familiarly Explained and Illustrated by Experiments

VOLUME 2:
ON COMPOUND BODIES

JANE HALDIMAND MARCET

CAMBRIDGE UNIVERSITY PRESS

Cambridge, New York, Melbourne, Madrid, Cape Town, Singapore,
São Paolo, Delhi, Dubai, Tokyo, Mexico City

Published in the United States of America by Cambridge University Press, New York

www.cambridge.org
Information on this title: www.cambridge.org/9781108016841

© in this compilation Cambridge University Press 2010

This edition first published 1817
This digitally printed version 2010

ISBN 978-1-108-01684-1 Paperback

This book reproduces the text of the original edition. The content and language reflect the beliefs, practices and terminology of their time, and have not been updated.

Cambridge University Press wishes to make clear that the book, unless originally published by Cambridge, is not being republished by, in association or collaboration with, or with the endorsement or approval of, the original publisher or its successors in title.

CONVERSATIONS

ON

CHEMISTRY;

IN WHICH

THE ELEMENTS OF THAT SCIENCE

ARE

FAMILIARLY EXPLAINED

AND

ILLUSTRATED BY EXPERIMENTS.

IN TWO VOLUMES.

The Fifth Edition, revised, corrected, and considerably enlarged.

VOL. II.

ON COMPOUND BODIES.

LONDON:
PRINTED FOR LONGMAN, HURST, REES, ORME, AND BROWN,
PATERNOSTER-ROW.
1817.

CONTENTS

OF

THE SECOND VOLUME.

ON COMPOUND BODIES.

CONVERSATION XIII.

ON THE ATTRACTION OF COMPOSITION.

Of the laws which regulate the Phenomena of the Attraction of Composition. — 1. It takes place only between Bodies of a different Nature. — 2. Between the most minute Particles only. — 3. Between 2, 3, 4, or more Bodies. — Of Compound or Neutral Salts. — 4. Produces a Change of Temperature. — 5. The Properties which characterise Bodies in their separate State, destroyed by Combination. — 6. The Force of Attraction estimated by that which is required by the Separation of the Constituents. — 7. Bodies have amongst themselves different Degrees of Attraction. — Of simple elective and double elective Attractions. — Of quiescent and divellent Forces. — Law of definite Proportions. — Decomposition of Salts by Voltaic Electricity.

CONVERSATION XIV.

Page

ON ALKALIES. 19

Of the Composition and general Properties of the Alkalies. — Of Potash. — Manner of preparing it — Pearl-ash. — Soap. — Carbonat of Potash. — Chemical Nomenclature. — Solution of Potash. — Of Glass. — Of Nitrat of Potash or Saltpetre. — Effect of Alkalies on Vegetable Colours. — Of Soda. — Of Ammonia or Volatile Alkali. — Muriat of Ammonia. — Ammoniacal Gas. — Composition of Ammonia — Hartshorn ad Sal Volatile. — Combustion of Ammoniacal Gas.

CONVERSATION XV.

ON EARTHS. 44

Composition of the Earths. — Of their Incombustibility. — Form the Basis of all Minerals. — Their Alkaline Properties. — Silex; its Properties and Uses in the Arts. — Alumine; its Uses in Pottery, &c. — Alkaline Earths. — Barytes. — Lime; its extensive chemical Properties and Uses in the Arts. — Magnesia. — Strontian.

CONVERSATION XVI.

ON ACIDS. 69

Nomenclature of the Acids. — Of the Classification of Acids. — 1st Class — Acids of simple and known Radicals, or Mineral Acids. — 2d Class — Acids of double Radicals, or Vegetable Acids. — 3d Class — Acids of triple Radicals or Animal Acids. — Of the Decomposition of Acids of the 1st Class by Combustible bodies.

CONVERSATION XVII.

Page

OF THE SULPHURIC AND PHOSPHORIC ACIDS: OR, THE COMBINATIONS OF OXYGEN WITH SULPHUR AND WITH PHOSPHORUS; AND OF THE SULPHATS AND PHOSPHATS. 80

Of the Sulphuric Acid. — Combustion of Animal or Vegetable Bodies by this Acid. — Method of preparing it. — The Sulphurous Acid obtained in the Form of Gas. —May be obtained from Sulphuric Acid.— May be reduced to Sulphur.—Is absorbable by Water.—Destroys Vegetable Colours. — Oxyd of Sulphur. — Of Salts in general. — Sulphats. — Sulphat of Potash, or Sal Polychrest.— Cold produced by the melting of Salts.— Sulphat of Soda, or Glauber's Salt. — Heat evolved during the Formation of Salts. — Crystallisation of Salts. — Water of Crystallisation. — Efflorescence and Deliquescence of Salts. — Sulphat of Lime, Gypsum or Plaister of Paris. — Sulphat of Magnesia. — Sulphat of Alumine, or Alum. — Sulphat of Iron. — Of Ink. — Of the Phosphoric and Phosphorous Acids. — Phosphorus obtained from Bones. — Phosphat of Lime.

CONVERSATION XVIII.

OF THE NITRIC AND CARBONIC ACIDS: OR THE COMBINATION OF OXYGEN WITH NITROGEN AND WITH CARBON; AND OF THE NITRATS AND CARBONATS. 100

Nitrogen susceptible of various Degrees of Acidification. — Of the Nitric Acid. — Its Nature and Composition discovered by Mr. Cavendish. — Obtained ftom Nitrat of Potash. — Aqua Fortis. — Nitric Acid may be con-

Page

verted into Nitrous Acid. — Nitric Oxyd Gas. — Its Conversion into Nitrous Acid Gas. — Used as an Eudiometrical Test. — Gaseous Oxyd of Nitrogen, or exhilarating Gas, obtained from Nitrat of Ammonia. — Its singular Effects on being respired. — Nitrats. — Of Nitrat of Potash, Nitre or Saltpetre. — Of Gunpowder. — Causes of Detonation. — Decomposition of Nitre. — Deflagration. — Nitrat of Ammonia. — Nitrat of Silver. — Of the Carbonic Acid. — Formed by the Combustion of Carbon.— Constitutes a component Part of the Atmosphere. — Exhaled in some Caverns. — Grotto del Cane. — Great Weight of this Gas. — Produced from calcareous Stones by Sulphuric Acid. — Deleterious Effects of this Gas when respired. — Sources which keep up a Supply of this Gas in the Atmosphere. — Its Effects on Vegetation.— Of the Carbonats of Lime; Marble, Chalk, Shells, Spars, and calcareous Stones.

CONVERSATION XIX.

ON THE BORACIC, FLUORIC, MURIATIC, AND OXYGENATED MURIATIC ACIDS; AND ON MURIATS. 131

On the Boracic Acid.— Its Decomposition by Sir H. Davy. — Its Basis Boracium. — Its Recomposition. — Its Uses in the Arts.— Borax or Borat of Soda.— Of the Fluoric Acid. — Obtained from Fluor; corrodes Siliceous Earth; its supposed Composition. — Fluorine; its supposed Basis. — Of the Muriatic Acid.— Obtained from Muriats.— Its gaseous Form.— Is absorbable by Water. — Its Decomposition.— Is susceptible of a stronger Degree of Oxygenation. — Oxygenated Muriatic Acid. — Its gaseous Form and other Properties. — Combustion of Bodies in this Gas. — It dissolves Gold. — Compo-

Page

sition of Aqua Regia. — Oxygenated Muriatic Acid destroys all Colours. — Sir H. Davy's Theory of the Nature of Muriatic and Oxymuriatic Acid. — Chlorine. Used for Bleaching and for Fumigations. — Its offensive Smell, &c. — Muriats. — Muriat of Soda, or common Salt. — Muriat of Ammonia. — Oxygenated Muriat of Potash. — Detonates with Sulphur, Phosphorus, &c. — Experiment of burning Phosphorus under Water by means of this Salt and of Sulphuric Acid,

CONVERSATION XX.

ON THE NATURE AND COMPOSITION OF VEGETABLES. 162

Of organised Bodies. — Of the Functions of Vegetables. — Of the Elements of Vegetables. — Of the Materials of Vegetables. — Analysis of Vegetables. — Of Sap. — Mucilage, or Gum. — Sugar. — Manna, and Honey. — Gluten. —Vegetable Oils. — Fixed Oils, Linseed, Nut, and Olive Oils. — Volatile Oils, forming Essences and Perfumes. — Camphor. — Resins and Varnishes. — Pitch, Tar, Copal, Mastic, &c. — Gum Resins. — Myrrh, Assafœtida, &c. — Caoutchouc, or Gum Elastic. — Extractive colouring Matter; its Use in the Arts of Dyeing and Painting. — Tannin; its Use in the Art of preparing Leather. — Woody Fibre. — Vegetable Acids. — The Alkalies and Salts contained in Vegetables.

CONVERSATION XXI.

ON THE DECOMPOSITION OF VEGETABLES. 202

Of Fermentation in general. — Of the Saccharine Fermentation, the Product of which is Sugar. — Of the Vinous Fermentation, the Product of which is Wine.

Page

—Alcohol, or Spirit of Wine. — Analysis of Wine by Distillation. — Of Brandy, Rum, Arrack, Gin, &c. — Tartrit of Potash, or Cream of Tartar. — Liqueurs. — Chemical Properties of Alcohol. — Its Combustion. — Of Ether. — Of the Acetous Fermentation, the Product of which is Vinegar. — Fermentation of Bread. — Of the Putrid Fermentation, which reduces Vegetables to their Elements. — Spontaneous Succession of these Fermentations. — Of Vegetables said to be petrified. — Of Bitumens: Naphtha, Asphaltum, Jet, Coal, Succin, or Yellow Amber.—Of Fossil Wood, Peat, and Turf.

CONVERSATION XXII.

HISTORY OF VEGETATION. 245

Connexion between the Vegetable and Animal Kingdoms.— Of Manures.— Of Agriculture.— Inexhaustible Sources of Materials for the Purposes of Agriculture. — Of sowing Seed. — Germination of the Seed. — Function of the Leaves of Plants. — Effects of Light and Air on Vegetation.— Effects of Water on Vegetation. — Effects of Vegetation on the Atmosphere. — Formation of Vegetable Materials by the Organs of Plants. — Vegetable Heat.— Of the Organs of Plants.— Of the Bark, consisting of Epidermis, Parenchyma, and Cortical Layers. — Of Alburnum, or Wood. — Leaves, Flowers, and Seeds.— Effects of the Season on Vegetation.—Vegetation of Evergreens in Winter.

CONVERSATION XXIII.

ON THE COMPOSITION OF ANIMALS. 276

Elements of Animals. — Of the principal Materials of Animals, viz. Gelatine, Albumen, Fibrine, Mucus. —

Page

Of Animal Acids. — Of Animal Colours, Prussian Blue, Carmine, and Ivory Black.

CONVERSATION XXIV.

ON THE ANIMAL ECONOMY. 297

Of the principal Animal Organs. — Of Bones, Teeth, Horns, Ligaments, and Cartilage. — Of the Muscles, constituting the Organs of Motion. — Of the Vascular System, for the Conveyance of Fluids. — Of the Glands, for the Secretion of Fluids. — Of the Nerves, constituting the Organs of Sensation. — Of the Cellular Substance which connects the several Organs. — Of the Skin.

CONVERSATION XXV.

ON ANIMALISATION, NUTRITION, AND RESPIRATION. 314

Digestion. — Solvent Power of the Gastric Juice. — Formation of a Chyle. — Its Assimilation, or Conversion into Blood. — Of Respiration. — Mechanical Process of Respiration. — Chemical Process of Respiration. — Of the Circulation of the Blood. — Of the Functions of the Arteries, the Veins, and the Heart. — Of the Lungs. — Effects of Respiration on the Blood.

CONVERSATION XXVI

ON NIMALAHEAT; AND OF VARIOUS ANIMAL PRODUCTS. 336

Of the Analogy of Combustion and Respiration. — Animal Heat evolved in the Lungs. — Animal Heat evolved in the Circulation. — Heat produced by Fever. — Perspi-

Page

ration. — Heat produced by Exercise. — Equal Temperature of Animals at all Seasons. — Power of the Animal Body to resist the Effects of Heat. — Cold produced by Perspiration. — Respiration of Fish and of Birds. — Effects of Respiration on Muscular Strength. — Of several Animal Products, viz. Milk, Butter, and Cheese; Spermaceti; Ambergris; Wax; Lac; Silk; Musk; Civet; Castor. — Of the putrid Fermentation. — Conclusion.

CONVERSATIONS

ON

CHEMISTRY.

CONVERSATION XIII.

ON THE ATTRACTION OF COMPOSITION.

MRS. B.

HAVING completed our examination of the simple or elementary bodies, we are now to proceed to those of a compound nature; but before we enter on this extensive subject, it will be necessary to make you acquainted with the principal laws by which chemical combinations are governed.

You recollect, I hope, what we formerly said of the nature of the attraction of composition, or chemical attraction, or affinity, as it is also called?

EMILY.

Yes, I think perfectly; it is the attraction that

subsists between bodies of a different nature, which occasions them to combine and form a compound, when they come in contact, and, according to Sir H. Davy's opinion, this effect is produced by the attraction of the opposite electricities, which prevail in bodies of different kinds.

MRS. B.

Very well; your definition comprehends the first law of chemical attraction, which is, that *it takes place only between bodies of a different nature;* as, for instance, between an acid and an alkali; between oxygen and a metal, &c.

CAROLINE.

That we understand of course; for the attraction between particles of a similar nature is that of aggregation, or cohesion, which is independent of any chemical power.

MRS. B.

The 2d law of chemical attraction is, that *it takes place only between the most minute particles of bodies;* therefore, the more you divide the particles of the bodies to be combined, the more readily they act upon each other.

CAROLINE.

That is again a circumstance which we might

have supposed, for the finer the particles of the two substances are, the more easily and perfectly they will come in contact with each other, which must greatly facilitate their union. It was for this purpose, you said, that you used iron filings, in preference to wires or pieces of iron, for the decomposition of water.

MRS. B.

It was once supposed that no mechanical power could divide bodies into particles sufficiently minute for them to act on each other; and that, in order to produce the extreme division requisite for a chemical action, one, if not both of the bodies, should be in a fluid state. There are, however, a few instances in which two solid bodies, very finely pulverized, exert a chemical action on one another; but such exceptions to the general rule are very rare indeed.

EMILY.

In all the combinations that we have hitherto seen, one of the constituents has, I believe, been either liquid or aëriform. In combustions, for instance, the oxygen is taken from the atmosphere, in which it existed in the state of gas; and whenever we have seen acids combine with metals or with alkalies, they were either in a liquid or an aëriform state.

MRS. B.

The 3d law of chemical attraction is, that *it can take place between two, three, four, or even a greater number of bodies.*

CAROLINE.

Oxyds and acids are bodies composed of two constituents; but I recollect no instance of the combination of a greater number of principles.

MRS. B.

The compound salts, formed by the union of the metals with acids, are composed of three principles. And there are salts formed by the combination of the alkalies with the earths which are of a similar description.

CAROLINE.

Are they of the same kind as the metallic salts?

MRS. B.

Yes; they are very analogous in their nature, although different in many of their properties.

A methodical nomenclature, similar to that of the acids, has been adopted for the compound salts. Each individual salt derives its name from its constituent parts, so that every name implies a knowledge of the composition of the salt.

The three alkalies, the alkaline earths, and the

metals, are called *salifiable bases* or *radicals;* and the acids, *salifying principles.* The name of each salt is composed both of that of the acid and the salifiable base; and it terminates in *at* or *it*, according to the degree of the oxygenation of the acid. Thus, for instance, all those salts which are formed by the combination of the sulphuric acid with any of the salifiable bases are called *sulphats*, and the name of the radical is added for the specific distinction of the salt; if it be potash, it will compose a *sulphat of potash;* if ammonia, *sulphat of ammonia*, &c.

EMILY.

The crystals which we obtained from the combination of iron and sulphuric acid were therefore *sulphat of iron?*

MRS. B.

Precisely; and those which we prepared by dissolving copper in nitric acid, *nitrat of copper*, and so on. — But this is not all; if the salt be formed by that class of acids which ends in *ous*, (which you know indicates a less degree of oxygenation,) the termination of the name of the salt will be in *it*, as *sulphit of potash*, *sulphit of ammonia*, &c.

EMILY.

There must be an immense number of compound

salts, since there is so great a variety of salifiable radicals, as well as of salifying principles.

MRS. B.

Their real number cannot be ascertained, since it increases every day. But we must not proceed further in the investigation of the compound salts, until we have completed the examination of the nature of the ingredients of which they are composed.

The 4th law of chemical attraction is, that *a change of temperature always takes place at the moment of combination.* This arises from the extrication of the two electricities in the form of caloric, which takes place when bodies unite; and also sometimes in part from a change of capacity of the bodies for heat, which always takes place when the combination is attended with an increase of density, but more especially when the compound passes from the liquid to the solid form. I shall now show you a striking instance of a change of temperature from chemical union, merely by pouring some nitrous acid on this small quantity of oil of turpentine — the oil will instantly combine with the oxygen of the acid, and produce a considerable change of temperature.

CAROLINE.

What a blaze! The temperature of the oil and

the acid must be greatly raised, indeed, to produce such a violent combustion.

MRS. B.

There is, however, a peculiarity in this combustion, which is, that the oxygen, instead of being derived from the atmosphere alone, is principally supplied by the acid itself.

EMILY.

And are not all combustions instances of the change of temperature produced by the chemical combination of two bodies?

MRS. B.

Undoubtedly; when oxygen loses its gaseous form, in order to combine with a solid body, it becomes condensed, and the caloric evolved produces the elevation of temperature. The specific gravity of bodies is at the same time altered by chemical combination; for in consequence of a change of capacity for heat, a change of density must be produced.

CAROLINE.

That was the case with the sulphuric acid and water, which, by being mixed together, gave out a great deal of heat, and increased in density.

MRS. B.

The 5th law of chemical attraction is, that *the properties which characterise bodies, when separate, are altered or destroyed by their combination.*

CAROLINE.

Certainly; what, for instance, can be so different from water as the hydrogen and oxygen gases?

EMILY.

Or what more unlike sulphat of iron than iron or sulphuric acid?

MRS. B.

Every chemical combination is an illustration of this rule. But let us proceed —

The 6th law is, that *the force of chemical affinity between the constituents of a body is estimated by that which is required for their separation.* This force is not always proportional to the facility with which bodies unite; for manganese, for instance, which, you know, is so much disposed to unite with oxygen that it is never found in a metallic state, yields it more easily than any other metal.

EMILY.

But, Mrs. B., you speak of estimating the force of attraction between bodies, by the force required to separate them; how can you measure these forces?

MRS. B.

They cannot be precisely *measured*, but they are comparatively ascertained by experiment, and can be represented by numbers which express the relative degrees of attraction.

The 7th law is, that *bodies have amongst themselves different degrees of attraction.* Upon this law, (which you may have discovered yourselves long since,) the whole science of chemistry depends; for it is by means of the various degrees of affinity which bodies have for each other, that all the chemical compositions and decompositions are effected. Every chemical fact or experiment is an instance of the same kind; and whenever the decomposition of a body is performed by the addition of any single new substance, it is said to be effected by *simple elective attractions.* But it often happens that no simple substance will decompose a body, and that, in order to effect this, you must offer to the compound a body which is itself composed of two, or sometimes three principles, which would not, each separately, perform the decomposition. In this case there are two new compounds formed in consequence of a reciprocal decomposition and recomposition. All instances of this kind are called *double elective attractions.*

CAROLINE.

I confess I do not understand this clearly.

MRS. B.

You will easily comprehend it by the assistance of this diagram, in which the reciprocal forces of attraction are represented by numbers:

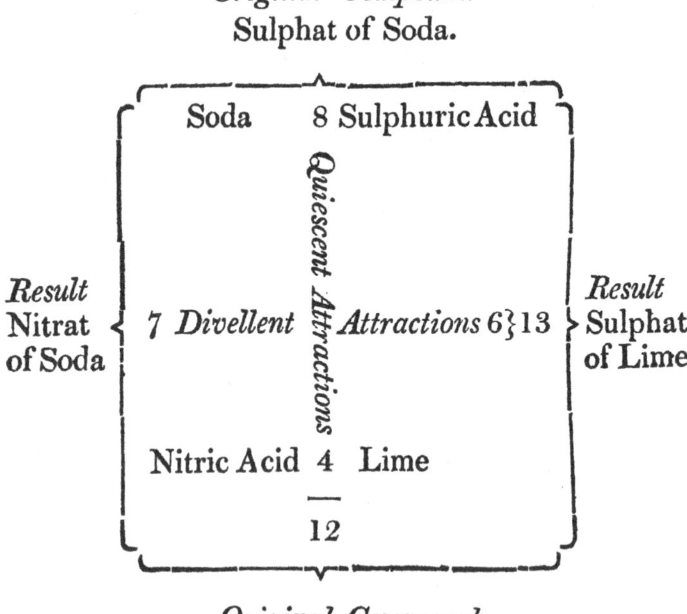

We here suppose that we are to decompose sulphat of soda; that is, to separate the acid from the alkali; if, for this purpose, we add some lime, in order to make it combine with the acid, we shall fail in our attempt, because the soda and the sulphuric acid attract each other by a force which is superior, and (by way of supposition) is represented by the number 8; while the lime tends to unite with this acid by an affinity equal only to the number 6. It is plain, therefore, that the sulphat of soda will not be decomposed, since a force equal to 8 cannot be overcome by a force equal only to 6.

CAROLINE.

So far, this appears very clear.

MRS. B.

If, on the other hand, we endeavour to decompose this salt by nitric acid, which tends to combine with soda, we shall be equally unsuccessful, as nitric acid tends to unite with the alkali by a force equal only to 7.

In neither of these cases of simple elective attraction, therefore, can we accomplish our purpose. But let us previously combine together the lime and nitric acid, so as to form a nitrat of lime, a compound salt, the constituents of which are united by a power equal to 4. If then we present this compound to the sulphat of soda, a decomposition will ensue, because the sum of the forces whidh tend to preserve the two salts in their actual state is not equal to that of the forces which tend to decompose them, and to form new combinations. The nitric acid, therefore, will combine with the soda, and the sulphuric acid with the lime.

CAROLINE.

I understand you now very well. This double effect takes place because the numbers 8 and 4, which represent the degrees of attraction of the constituents of the two original salts, make a sum less than the numbers 7 and 6, which represent the degrees of attraction of the two new compounds that will in consequence be formed.

MRS. B.

Precisely so.

CAROLINE.

But what is the meaning of *quiescent* and *divellent* forces, which are written in the diagram?

MRS. B.

Quiescent forces are those which tend to preserve compounds in a state of rest, or such as they actually are: divellent forces, those which tend to destroy that state of combination, and to form new compounds.

These are the principal circumstances relative to the doctrine of chemical attractions, which have been laid down as rules by modern chemists; a few others might be mentioned respecting the same theory, but of less importance, and such as would take us too far from our plan. I should, however, not omit to mention that Mr. Berthollet, a celebrated French chemist, has questioned the uniform operation of elective attraction, and has advanced the opinion, that, in chemical combinations, the changes which take place depend not only upon the affinities, but also, in some degree, on the respective quantities of the substances concerned, on the heat applied during the process, and some other circumstances.

CAROLINE.

In that case, I suppose, there would hardly be two compounds exactly similar, though composed of the same materials?

MRS. B.

On the contrary, it is found that a remarkable uniformity prevails, as to proportions, between the ingredients of bodies of similar composition. Thus water, as you may recollect to have seen in a former conversation, is composed of two volumes of hydrogen gas to one of oxygen, and this is always found to be precisely the proportion of its constituents, from whatever source the water be derived. The same uniformity prevails with regard to the various salts; the acid and alkali, in each kind of salt, being always found to combine in the same proportions. Sometimes, it is true, the same acid, and the same alkali, are capable of making two distinct kinds of salts; but in all these cases it is found that one of the salts contains just twice, or in some instances, thrice as much acid, or alkali, as the other.

EMILY.

If the proportions in which bodies combine are so constant and so well defined, how can Mr. Berthollet's remark be reconciled with this uniform system of combination?

MRS. B.

Great as that philosopher's authority is in chemistry, it is now generally supposed that his doubts on this subject were in a great degree groundless, and that the exceptions he has observed in the laws of definite proportions, have been only apparent, and may be accounted for consistently with those laws.

CAROLINE.

Pray, Mrs. B., can you decompose a salt by means of electricity, in the same way as we decompose water?

MRS. B.

Undoubtedly; and I am glad this question occurred to you, because it gives me an opportunity of showing you some very interesting experiments on the subject.

If we dissolve a quantity, however small, of any salt in a glass of water, and if we plunge into it the extremities of the wires which proceed from the two ends of the Voltaic battery, the salt will be gradually decomposed, the acid being attracted by the positive, and the alkali by the negative wire.

EMILY.

But how can you render that decomposition perceptible.

MRS. B.

By placing in contact with the extremities of each wire, in the solution, pieces of paper stained with certain vegetable colours, which are altered by the contact of an acid or an alkali. Thus this blue vegetable preparation called litmus becomes red when touched by an acid; and the juice of violets becomes green by the contact of an alkali.

But the experiment can be made in a much more distinct manner, by receiving the extremities of the wires into two different vessels, so that the alkali shall appear in one vessel and the acid in the other.

CAROLINE.

But then the Voltaic circle will not be completed; how can any effect be produced?

MRS. B.

You are right; I ought to have added that the two vessels must be connected together by some interposed substance capable of conducting electricity. A piece of moistened cotton-wick answers this purpose very well. You see that the cotton (PLATE XIII. fig. 2. c.) has one end immersed in one glass and the other end in the other, so as to establish a communication between any fluids contained in them. We shall now put into each of the glasses a little glauber salt, or sulphat of soda, (which consists of an acid and an alkali,) and then we shall fill the glasses with water, which will dissolve the salt. Let us now connect the glasses by means of the wires (e, d,) with the two ends of the battery, thus . . . .

CAROLINE.

The wires are already giving out small bubbles; is this owing to the decomposition of the salt?

MRS. B.

No; these are bubbles produced by the decomposition of the water, as you saw in a former experiment. In order to render the separation of the acid from the alkali visible, I pour into the glass (a), which is connected with the positive wire, a few drops of a solution of litmus,

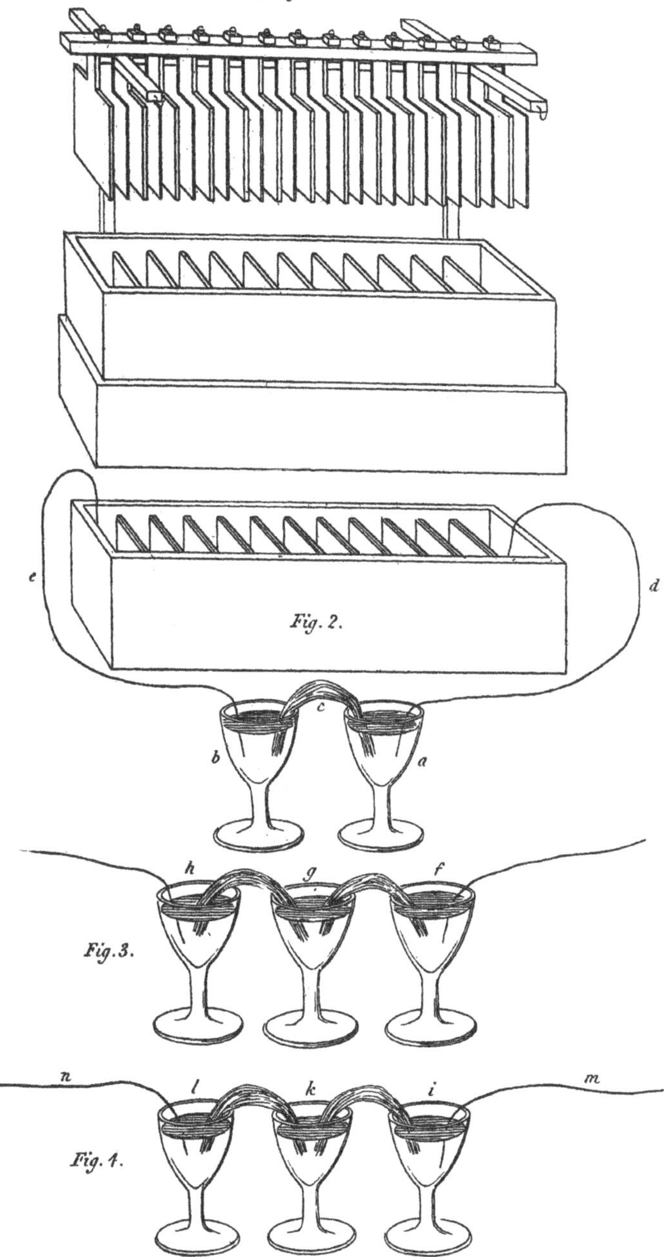

Fig. 1. Voltaic Battery of improved construction with the Plates out of the Cells.
Fig. 2. 3 & 4. Instances of Chemical decomposition by the Voltaic Battery.

Lowry sculp.

which the least quantity of acid turns red; and into the other glass (b), which is connected with the negative wire, I pour a few drops of the juice of violets . . . .

EMILY.

The blue solution is already turning red all round the wire.

CAROLINE.

And the violet solution is beginning to turn green. This is indeed very singular!

MRS. B.

You will be still more astonished when we vary the experiment in this manner: — These three glasses (fig. 3. f, g, h,) are, as in the former instance, connected together by wetted cotton, but the middle one alone contains a saline solution, the two others containing only distilled water, coloured as before by vegetable infusions. Yet, on making the connection with the battery, the alkali will appear in the negative glass (h), and the acid in the positive glass (f), though neither of them contained any saline matter.

EMILY.

So that the acid and alkali must be conveyed right and left from the central glass, into the other glasses, by means of the connecting moistened cotton?

MRS. B.

Exactly so; and you may render the experiment still more striking, by putting into the central glass (k, fig. 3.) an alkaline solution, the glauber salt being placed into the negative glass (l), and the positive glass (i) containing only water. The acid will be attracted by the positive wire (m), and will actually appear in the vessel (i), after passing through the alkaline solution (k), without combining with it, although, you know, acids and alkalies are so much disposed to combine. — But this conversation has already much exceeded our usual limits, and we cannot enlarge more upon this interesting subject at present.

CONVERSATION XIV.

ON ALKALIES.

MRS. B.

HAVING now given you some idea of the laws by which chemical attractions are governed, we may proceed to the examination of bodies which are formed in consequence of these attractions.

The first class of compounds that present themselves to our notice, in our gradual ascent to the most complicated combinations, are bodies composed of only two principles. The sulphurets, phosphurets, carburets, &c. are of this description; but the most numerous and important of these compounds are the combinations of oxygen with the various simple substances with which it has a tendency to unite. Of these you have already acquired some knowledge, but it will be necessary to enter into further particulars respecting the nature and properties of those most deserving our notice. Of this class are the ALKA-

LIES and the EARTHS, which we shall successively examine.

We shall first take a view of the alkalies, of which there are three, viz. POTASH, SODA, and AMMONIA. The two first are called *fixed alkalies*, because they exist in a solid form at the temperature of the atmosphere, and require a great heat to be volatilised. They consist, as you already know, of metallic bases combined with oxygen. In potash, the proportions are about eighty-six parts of potassium to fourteen of oxygen; and in soda, seventy-seven parts of sodium to twenty three of oxygen. The third alkali, ammonia, has been distinguished by the name of *volatile alkali*, because its natural form is that of gas. Its composition is of a more complicated nature, of which we shall speak hereafter.

Some of the earths bear so strong a resemblance in their properties to the alkalies, that it is difficult to know under which head to place them. The celebrated French chemist, Fourcroy, has classed two of them (barytes and strontites) with the alkalies; but as lime and magnesia have almost an equal title to that rank, I think it better not to separate them, and therefore have adopted the common method of classing them with the earths, and of distinguishing them by the name of *alkaline earths.*

The general properties of alkalies are, an acrid

burning taste, a pungent smell, and a caustic action on the skin and flesh.

CAROLINE.

I wonder they should be caustic, Mrs. B., since they contain so little oxygen.

MRS. B.

Whatever substance has an affinity for any one of the constituents of animal matter, sufficiently powerful to decompose it, is entitled to the appellation of caustic. The alkalies, in their pure state, have a very strong attraction for water, for hydrogen, and for carbon, which, you know, are the constituent principles of oil, and it is chiefly by absorbing these substances from animal matter that they effect its decomposition; for, when diluted with a sufficient quantity of water, or combined with any oily substance, they lose their causticity.

But, to return to the general properties of alkalies — they change, as we have already seen, the colour of syrup of violets, and other blue vegetable infusions, to green; and have, in general, a very great tendency to unite with acids, although the respective qualities of these two classes of bodies form a remarkable contrast.

We shall examine the result of the combination of acids and alkalies more particularly hereafter.

It will be sufficient at present to inform you, that whenever acids are brought in contact with alkalies, or alkaline earths, they unite with a remarkable eagerness, and form compounds perfectly different from either of their constituents; these bodies are called *neutral* or *compound salts.*

The dry white powder which you see in this phial is pure caustic POTASH; it is very difficult to preserve it in this state, as it attracts, with extreme avidity, the moisture from the atmosphere, and if the air were not perfectly excluded, it would, in a very short time, be actually melted.

EMILY.

It is then, I suppose, always found in a liquid state?

MRS. B.

No; it exists in nature in a great variety of forms and combinations, but is never found in its pure separate state; it is combined with carbonic acid, with which it exists in every part of the vegetable kingdom, and is most commonly obtained from the ashes of vegetables, which are the residue that remains after all the other parts have been volatilised by combustion.

CAROLINE.

But you once said, that after all the volatile

parts of a vegetable were evaporated, the substance that remained was charcoal?

MRS. B.

I am surprised that you should still confound the processes of volatilisation and combustion. In order to procure charcoal, we evaporate such parts as can be reduced to vapour by the operation of heat alone; but when we *burn* the vegetable, we burn the carbon also, and convert it into carbonic acid gas.

CAROLINE.

That is true; I hope I shall make no more mistakes in my favourite theory of combustion.

MRS. B.

Potash derives its name from the *pots* in which the vegetables, from which it was obtained, used formerly to be burnt; the alkali remained mixed with the ashes at the bottom, and was thence called potash.

EMILY.

The ashes of a wood-fire, then, are potash, since they are vegetable ashes?

MRS. B.

They always contain more or less potash, but are very far from consisting of that substance alone, as they are a mixture of various earths and salts which remain after the combustion of vegetables, and from which it is not easy to separate the alkali in its pure form. The process by which potash is obtained, even in the imperfect state in which it is used in the arts, is much more complicated than simple combustion. It was once deemed impossible to separate it entirely from all foreign substances, and it is only in chemical laboratories that it is to be met with in the state of purity in which you find it in this phial. Wood-ashes are, however, valuable for the alkali which they contain, and are used for some purposes without any further preparation. Purified in a certain degree, they make what is commonly called *pearlash*, which is of great efficacy in taking out grease, in washing linen, &c.; for potash combines readily with oil or fat, with which it forms a compound well known to you under the name of *soap*.

CAROLINE.

Really! Then I should think it would be better to wash all linen with pearlash than with soap, as, in the latter case, the alkali being already combined with oil, must be less efficacious in extracting grease.

MRS. B.

Its effect would be too powerful on fine linen, and would injure its texture; pearlash is therefore only used for that which is of a strong coarse kind. For the same reason you cannot wash your hands with plain potash; but, when mixed with oil in the form of soap, it is soft as well as cleansing, and is therefore much better adapted to the purpose.

Caustic potash, as we already observed, acts on the skin, and animal fibre, in virtue of its attraction for water and oil, and converts all animal matter into a kind of saponaceous jelly.

EMILY.

Are vegetables the only source from which potash can be derived?

MRS. B.

No: for though far most abundant in vegetables, it is by no means confined to that class of bodies, being found also on the surface of the earth, mixed with various minerals, especially with earths and stones, whence it is supposed to be conveyed into vegetables by the roots of the plant. It is also met with, though in very small quantities, in some animal substances. The most common state of potash is that of *carbonat*; I suppose you understand what that is?

EMILY.

I believe so; though I do not recollect that you ever mentioned the word before. If I am not mistaken, it must be a compound salt, formed by the union of carbonic acid with potash.

MRS. B.

Very true; you see how admirably the nomenclature of modern chemistry is adapted to assist the memory; when you hear the name of a compound, you necessarily learn what are its constituent parts; and when you are acquainted with these constituents, you can immediately name the compound which they form.

CAROLINE.

Pray, how were bodies arranged and distinguished before this nomenclature was introduced?

MRS. B.

Chemistry was then a much more difficult study; for every substance had an arbitrary name, which it derived either from the person who discovered it, as *Glauber's salts* for instance; or from some other circumstance relative to it, though quite unconnected with its real nature, as *potash.*

These names have been retained for some of the simple bodies; for as this class is not numerous, and therefore can easily be remembered, it has not been thought necessary to change them.

EMILY.

Yet I think it would have rendered the new nomenclature more complete to have methodised the names of the elementary, as well as of the compound bodies, though it could not have been done in the same manner. But the names of the simple substances might have indicated their nature, or, at least, some of their principal properties; and if, like the acids and compound salts, all the simple bodies had a similar termination, they would have been immediately known as such. So complete and regular a nomenclature would, I think, have given a clearer and more comprehensive view of chemistry than the present, which is a medley of the old and new terms.

MRS. B.

But you are not aware of the difficulty of introducing into science an entire set of new terms; it obliges all the teachers and professors to go to school again, and if some of the old names, that are least exceptionable, were not left as an introduction to the new ones, few people would have had industry and perseverance enough to submit to the study of a completely new language; and the inferior classes of artists, who can only act from habit and routine, would, at least for a time, have felt material inconvenience from a total change of their habitual terms. From these considerations, Lavoisier and his colleagues, who invented the new nomenclature, thought it most prudent to leave a few links of the old chain, in order to connect it with the new one. Besides, you may easily conceive the inconvenience which might arise from giving a regular nomenclature to substances, the simple nature of which is always uncertain; for the new names might, perhaps, have proved to have been founded in error. And, indeed, cautious as the inventors of the modern chemical language have been, it has already been found necessary to modify it in many respects. In those few cases, however, in which new terms have been adopted to designate simple bodies, these names have been so contrived as to indicate one of the chief properties of the body in question; this is the case with oxygen, which, as I explained to you, signifies generator of acids; and hydrogen generator of water. If all the elementary bodies had a similar termination, as you propose, it would be necessary to change the name of any that might hereafter be found of a compound nature, which would be very inconvenient in this age of discovery.

But to return to the alkalies. — We shall now try to melt some of this caustic potash in a little water, as a circumstance occurs during its solution very worthy of observation. — Do you feel the heat that is produced?

CAROLINE.

Yes, I do; but is not this directly contrary to

our theory of latent heat, according to which hea is disengaged when fluids become solid, and cold produced when solids are melted?

MRS. B.

The latter is really the case in all solutions; and if the solution of caustic alkalies seems to make an exception to the rule, it does not, I believe, form any solid objection to the theory. The matter may be explained thus: When water first comes in contact with the potash, it produces an effect similar to the slaking of lime, that is, the water is solidified in combining with the potash, and thus loses its latent heat; this is the heat that you now feel, and which is, therefore, produced not by the melting of the solid, but by the solidification of the fluid. But when there is more water than the potash can absorb and solidify, the latter then yields to the solvent power of the water; and if we do not perceive the cold produced by its melting, it is because it is counterbalanced by the heat previously disengaged.

A very remarkable property of potash is the

* This defence of the general theory, however plausible, is liable to some obvious objections. The phenomenon might perhaps be better accounted for by supposing that a solution of alkali in water has less capacity for heat than either water or alkali in their separate state.

formation of glass by its fusion with siliceous earth. You are not yet acquainted with this last substance, further than its being in the list of simple bodies. It is sufficient, for the present, that you should know that sand and flint are chiefly composed of it; alone, it is infusible, but mixed with potash, it melts when exposed to the heat of a furnace, combines with the alkali, and runs into glass.

CAROLINE.

Who would ever have supposed that the same substance which converts transparent oil into such an opake body as soap, should transform that opake substance, sand, into transparent glass!

MRS. B.

The transparency, or opacity of bodies, does not, I conceive, depend so much upon their intimate nature, as upon the arrangement of their particles: we cannot have a more striking instance of this, than is afforded by the different states of carbon, which, though it commonly appears in the form of a black opake body, sometimes assumes the most dazzling transparent form in nature, that of diamond, which, you recollect, is carbon, and which, in all probability, derives its beautiful transparency from the peculiar arrangement of its particles during their crystallisation.

EMILY.

I never should have supposed that the formation of glass was so simple a process as you describe it.

MRS. B.

It is by no means an easy operation to make perfect glass; for if the sand, or flint, from which the siliceous earth is obtained, be mixed with any metallic particles, or other substance, which cannot be vitrified, the glass will be discoloured, or defaced, by opake specks.

CAROLINE.

That, I suppose, is the reason why objects so often appear irregular and shapeless through a common glass-window.

MRS. B.

This species of imperfection proceeds, I believe, from another cause. It is extremely difficult to prevent the lower part of the vessels, in which the materials of glass are fused, from containing a more dense vitreous matter than the upper, on account of the heavier ingredients falling to the bottom. When this happens, it occasions the appearance of veins or waves in the glass, from the difference of density in its several parts, which produces an irregular refraction of the rays of light that pass through it.

Another speeies of imperfection sometimes arise from the fusion not being continued for a length of time sufficient to combine the two ingredients completely, or from the due proportion of potash and silex (which are as two to one) not being carefully observed; the glass, in those cases, will be liable to alteration from the action of the air, of salts, and especially of acids, which will effect its decomposition by combining with the potash, and forming compound salts.

EMILY.

What an extremely useful substance potash is!

MRS. B.

Besides the great importance of potash in the manufactures of glass and soap, it is of very considerable utility in many of the other arts, and in its combinations with several acids, particularly the nitric, with which it forms saltpetre.

CAROLINE.

Then saltpetre must be a *nitrat of potash?* But we are not yet acquainted with the nitric acid?

MRS. B.

We shall therefore defer entering into the particulars of these combinations till we come to a general review of the compound salts. In order to avoid confusion, it will be better at present to confine ourselves to the alkalies.

EMILY.

Cannot you show us the change of colour which you said the alkalies produced on blue vegetable infusions?

MRS. B.

Yes; very easily. I shall dip a piece of white paper into this syrup of violets, which, you see, is of a deep blue, and dyes the paper of the same colour.—As soon as it is dry, we shall dip it into a solution of potash, which, though itself colourless, will turn the paper green—

CAROLINE.

So it has, indeed! And do the other alkalies produce a similar effect?

MRS. B.

Exactly the same.—We may now proceed to SODA, which, however important, will detain us but a very short time; as in all its general properties it very strongly resembles potash; indeed, so great is their similitude, that they have been long confounded, and they can now scarcely be distinguished, except by the difference of the salts which they form with acids.

The great source of this alkali is the sea, where,

combined with a peculiar acid, it forms the salt with which the waters of the ocean are so strongly impregnated.

EMILY.

Is not that the common table salt?

MRS. B.

The very same; but again we must postpone entering into the particulars of this interesting combination, till we treat of the neutral salts. Soda may be obtained from common salt; but the easiest and most usual method of procuring it is by the combustion of marine plants, an operation perfectly analogous to that by which potash is obtained from vegetables.

EMILY.

From what does soda derive its name?

MRS. B.

From a plant called by us *soda*, and by the Arabs *kali*, which affords it in great abundance. Kali has, indeed, given its name to the alkalies in general.

CAROLINE.

Does soda form glass and soap in the same manner as potash?

MRS. B.

Yes, it does; it is of equal importance in the

arts, and is even preferred to potash for some purposes; but you will not be able to distinguish their properties till we examine the compound salts which they form with acids; we must therefore leave soda for the present, and proceed to AMMONIA, or the VOLATILE ALKALI.

EMILY.

I long to hear something of this alkali; is it not of the same nature as hartshorn?

MRS. B.

Yes, it is, as you will see by-and-bye. This alkali is seldom found in nature in its pure state; it is most commonly extracted from a compound salt, called *sal ammoniac*, which was formerly imported from *Ammonia*, a region of Libya, from which both these salts and the alkali derive their names. The crystals contained in this bottle are specimens of this salt, which consists of a combination of ammonia and muriatic acid.

CAROLINE.

Then it should be called *muriat of ammonia;* for though I am ignorant what muriatic acid is, yet I know that its combination with ammonia cannot but be so called; and I am surprised to see sal ammoniac inscribed on the label.

MRS. B.

That is the name by which it has been so long known, that the modern chemists have not yet succeeded in banishing it altogether; and it is still sold under that name by druggists, though by scientific chemists it is more properly called muriat of ammonia.

CAROLINE.

Both the popular and the common name should be inscribed on labels — this would soon introduce the new nomenclature.

EMILY.

By what means can the ammonia be separated from the muriatic acid?

MRS. B.

By chemical attractions; but this operation is too complicated for you to understand, till you are better acquainted with the agency of affinities.

EMILY.

And when extracted from the salt, what kind of substance is ammonia?

MRS. B.

Its natural form, at the temperature of the atmosphere, when free from combination, is that of gas; and in this state it is called *ammoniacal gas*.

But it mixes very readily with water, and can be thus obtained in a liquid form.

CAROLINE.

You said that ammonia was more complicated in its composition than the other alkalies; pray of what principles does it consist?

MRS. B.

It was discovered a few years since, by Berthollet, a celebrated French chemist, that it consisted of about one part of hydrogen to four parts of nitrogen. Having heated ammoniacal gas under a receiver, by causing the electrical spark to pass repeatedly through it, he found that it increased considerably in bulk, lost all its alkaline properties, and was actually converted into hydrogen and nitrogen gases; and from the latest and most accurate experiments, the proportions appear to be, one volume of nitrogen gas to three of oxygen gas.

CAROLINE.

Ammonia, therefore, has not, like the two other alkalies, a metallic basis?

MRS. B.

It is believed it has, though it is extremely difficult to reconcile that idea with what I have just stated of its chemical nature. But the fact is, that although this supposed metallic basis of ammonia has never been obtained distinct and separate, yet both Professor Berzelius, of Stockholm, and Sir H. Davy, have succeeded in forming a combination of mercury with the basis of ammonia, which has so much the appearance of an amalgam, that it strongly corroborates the idea of ammonia having a metallic basis.* But these theoretical points are full of difficulties and doubts, and it would be useless to dwell any longer upon them.

Let us therefore return to the properties of volatile alkali. Ammoniacal gas is considerably lighter than oxygen gas, and only about half the weight of atmospherical air. It possesses most of the properties of the fixed alkalies; but cannot be of so much use in the arts on account of its volatile nature. It is, therefore, never employed in the manufacture of glass, but it forms soap with oils equally as well as potash and soda; it resembles them likewise in its strong attraction for water; for which reason it can be collected in a receiver over mercury only.

* This amalgam is easily obtained, by placing a globule of mercury upon a piece of muriat, or carbonat of ammonia, and electrifying this globule by the Voltaic battery. The globule instantly begins to expand to three or four times its former size, and becomes much less fluid, though without losing its metallic lustre, a change which is ascribed to the metallic basis of ammonia uniting with the mercury. This is an extremely curious experiment.

CAROLINE.

I do not understand this?

MRS. B.

Do you recollect the method which we used to collect gases in a glass-receiver over water?

CAROLINE.

Perfectly.

MRS. B.

Ammoniacal gas has so strong a tendency to unite with water, that, instead of passing through that fluid, it would be instantaneously absorbed by it. We can therefore neither use water for that purpose, nor any other liquid of which water is a component part; so that, in order to collect this gas, we are obliged to have recourse to mercury, (a liquid which has no action upon it,) and a mercurial bath is used instead of a water bath, such as we employed on former occasions. Water impregnated with this gas is nothing more than the fluid which you mentioned at the beginning of the conversation — hartshorn; it is the ammoniacal gas escaping from the water which gives it so powerful a smell.

EMILY.

But there is no appearance of effervescence in hartshorn.

MRS. B.

Because the particles of gas that rise from the water are too subtle and minute for their effect to be visible.

Water diminishes in density, by being impregnated with ammoniacal gas; and this augmentation of bulk increases its capacity for caloric.

EMILY.

In making hartshorn, then, or impregnating water with ammonia, heat must be absorbed, and cold produced?

MRS. B.

That effect would take place if it was not counteracted by another circumstance; the gas is liquefied by incorporating with the water, and gives out its latent heat. The condensation of the gas more than counterbalances the expansion of the water; therefore, upon the whole, heat is produced. — But if you dissolve ammoniacal gas with ice or snow, cold is produced. — Can you account for that?

EMILY.

The gas, in being condensed into a liquid, must give out heat; and, on the other hand, the snow or ice, in being rarefied into a liquid, must absorb heat; so that, between the opposite effects, I should have supposed the original temperature would have been preserved.

MRS. B.

But you have forgotten to take into the account the rarefaction of the water (or melted ice) by the impregnation of the gas; and this is the cause of the cold which is ultimately produced.

CAROLINE.

Is the *sal volatile* (the smell of which so strongly resembles hartshorn) likewise a preparation of ammonia?

MRS. B.

It is carbonat of ammonia dissolved in water; and which, in its concrete state, is commonly called salts of hartshorn. Ammonia is caustic, like the fixed alkalies, as you may judge by the pungent effects of hartshorn, which cannot be taken internally, nor applied to delicate external parts, without being plentifully diluted with water. — Oil and acids are very excellent antidotes for alkaline poisons; can you guess why?

CAROLINE.

Perhaps, because the oil combines with the alkali, and forms soap, and thus destroys its caustic properties; and the acid converts it into a compound salt, which, I suppose, is not so pernicious as caustic alkali.

MRS. B.

Precisely so.

Ammoniacal gas, if it be mixed with atmospherical air, and a burning taper repeatedly plunged into it, will burn with a large flame of a peculiar yellow colour.

EMILY.

But pray tell me, can ammonia be procured from this Lybian salt only?

MRS. B.

So far from it, that it is contained in, and may be extracted from, all animal substances whatever. Hydrogen and nitrogen are two of the chief constituents of animal matter; it is therefore not surprising that they should occasionally meet and combine in those proportions that compose ammonia. But this alkali is more frequently generated by the spontaneous decomposition of animal substances; the hydrogen and nitrogen gases that arise from putrefied bodies combine, and form the volatile alkali.

Muriat of ammonia, instead of being exclusively brought from Lybia, as it originally was, is now chiefly prepared in Europe, by chemical processes. Ammonia, although principally extracted from this salt, can also be produced by a great variety of other substances. The horns of cattle, especially those of deer, yield it in abundance, and it is from this circumstance that a solution of ammonia in water has been called hartshorn. It may likewise be procured from wool, flesh, and bones; in a word, any animal substance whatever yields it by decomposition.

We shall now lay aside the alkalies, however important the subject may be, till we treat of their combination with acids. The next time we meet we shall examine the earths.

CONVERSATION XV.

ON EARTHS.

MRS. B.

THE EARTHS, which we are to-day to examine, are nine in number:

SILEX,	STRONTITES,
ALUMINE,	YTTRIA,
BARYTES,	GLUCINA,
LIME,	ZIRCONIA.
MAGNESIA,	

The last three are of late discovery; their properties are but imperfectly known; and, as they have not yet been applied to use, it will be unnecessary to enter into any particulars respecting them; we shall confine our remarks, therefore, to the first five. They are composed, as you have already learnt, of a metallic basis combined with oxygen; and, from this circumstance, are incombustible.

CAROLINE.

Yet I have seen turf burnt in the country, and it makes an excellent fire; the earth becomes red hot, and produces a very great quantity of heat.

MRS. B.

It is not the earth that burns, my dear, but the roots, grass, and other remnants of vegetables that are intermixed with it. The caloric, which is produced by the combustion of these substances, makes the earth red hot, and this being a bad conductor of heat, retains its caloric a long time; but were you to examine it when cooled, you would find that it had not absorbed one particle of oxygen, nor suffered any alteration from the fire. Earth is, however, from the circumstance just mentioned, an excellent radiator of heat, and owes its utility, when mixed with fuel, solely to that property. It is in this point of view that Count Rumford has recommended balls of incombustible substances to be arranged in fire-places, and mixed with the coals, by which means the caloric disengaged by the combustion of the latter is more perfectly reflected into the room, and an expense of fuel is saved.

EMILY.

I expected that the list of earths would be much more considerable. When I think of the great variety of soils, I am astonished that there is not a greater number of earths to form them.

MRS. B.

You might, indeed, almost confine that number to four; for barytes, strontites, and the others of late discovery, act but so small a part in this great theatre, that they cannot be reckoned as essential to the general formation of the globe. And you must not confine your idea of earths to the formation of soil; for rock, marble, chalk, slate, sand, flint, and all kinds of stones, from the precious jewels to the commonest pebbles; in a word, all the immense variety of mineral products, may be referred to some of these earths, either in a simple state, or combined the one with the other, or blended with other ingredients.

CAROLINE.

Precious stones composed of earth! That seems very difficult to conceive.

EMILY.

Is it more extraordinary than that the most precious of all jewels, diamond, should be composed of carbon? But diamond forms an exception, Mrs. B.; for, though a stone, it is not composed of earth.

MRS. B.

I did not specify the exception, as I knew you were so well acquainted with it. Besides, I would call a diamond a mineral rather than a stone, as the latter term always implies the presence of some earth.

CAROLINE.

I cannot conceive how such coarse materials can be converted into such beautiful productions.

MRS. B.

We are very far from understanding all the secret resources of nature; but I do not think the spontaneous formation of the crystals, which we call precious stones, one of the most difficult phenomena to comprehend.

By the slow and regular work of ages, perhaps of hundreds of ages, these earths may be gradually dissolved by water, and as gradually deposited by their solvent in the undisturbed process of crystallisation. The regular arrangement of their particles, during their reunion in a solid mass, gives them that brilliancy, transparency, and beauty, for which they are so much admired; and renders them in appearance so totally different from their rude and primitive ingredients.

CAROLINE.

But how does it happen that they are spontaneously dissolved, and afterwards crystallised?

MRS. B.

The scarcity of many kinds of crystals, as rubies, emeralds, topazes, &c. shows that their formation is not an operation very easily carried on in nature. But cannot you imagine that when water, holding in solution some particles of earth, filters through the crevices of hills or mountains, and at length dribbles into some cavern, each successive drop may be slowly evaporated, leaving behind it the particle of earth which it held in solution? You know that crystallisation is more regular and perfect, in proportion as the evaporation of the solvent is slow and uniform; nature, therefore, who knows no limit of time, has, in all works of this kind, an infinite advantage over any artist who attempts to imitate such productions.

EMILY.

I can now conceive that the arrangement of the particles of earth, during crystallisation, may be such as to occasion transparency, by admitting a free passage to the rays of light; but I cannot understand why crystallised earths should assume such beautiful colours as most of them do. Sapphire, for instance, is of a celestial blue; ruby, a deep red; topaz, a brilliant yellow?

MRS. B.

Nothing is more simple than to suppose that the arrangement of their particles is such, as to transmit some of the coloured rays of light, and to reflect others, in which case the stone must appear of the colour of the rays which it reflects. But besides, it frequently happens that the colour of a stone is owing to a mixture of some metallic matter.

CAROLINE.

Pray, are the different kinds of precious stones each composed of one individual earth, or are they formed of a combination of several earths?

MRS. B.

A great variety of materials enters into the composition of most of them; not only séveral earths, but sometimes salts and metals. The earths, however, in their simple state, frequently form very beautiful crystals; and, indeed, it is in that state only that they can be obtained perfectly pure.

EMILY.

Is not the Derbyshire spar produced by the crystallisation of earths, in the way you have just explained? I have been in some of the subterraneous caverns where it is found, which are similar to those you have described.

MRS. B.

Yes; but this spar is a very imperfect specimen of crystallisation; it consists of a variety of ingredients confusedly blended together, as you may judge by its opacity, and by the various colours and appearances which it exhibits.

But, in examining the earths in their most perfect and agreeable form, we must not lose sight of that state in which they are commonly found, and which, if less pleasing to the eye, is far more interesting by its utility.

All the earths are more or less endowed with alkaline properties; but there are four, barytes, magnesia, lime, and strontites, which are called *alkaline earths*, because they possess those qualities in so great a degree, as to entitle them, in most respects, to the rank of alkalies. They combine and form compound salts with acids, in the same way as alkalies; they are, like them, susceptible of a considerable degree of causticity, and are acted upon in a similar manner by chemical tests. — The remaining earths, silex and alumine, with one or two others of late discovery, are in some degree more earthy, that is to say, they possess more completely the properties common to all the earths, which are, insipidity, dryness, unalterableness in the fire, infusibility, &c.

CAROLINE.

Yet, did you not tell us that silex, or siliceous earth, when mixed with an alkali, was fusible, and run into glass?

MRS. B.

Yes, my dear; but the characteristic properties of earths, which I have mentioned, are to be considered as belonging to them in a state of purity only; a state in which they are very seldom to be met with in nature. — Besides these general properties, each earth has its own specific characters, by which it is distinguished from any other substance. — Let us therefore review them separately.

SILEX, or SILICA, abounds in flint, sand, sandstone, agate, jasper, &c.; it forms the basis of many precious stones, and particularly of those which strike fire with steel. It is rough to the touch, scratches and wears away metals; it is acted upon by no acid but the fluoric, and is not soluble in water by any known process; but nature certainly dissolves it by means with which we are unacquainted, and thus produces a variety of siliceous crystals, and amongst these *rock crystal*, which is the purest specimen of this earth. Silex appears to have been intended by Providence to form the solid basis of the globe, to serve as a foundation for the original mountains, and give them that hardness and durability which has enabled them to resist the various revolutions which the surface of the earth has successively undergone. From these mountains siliceous rocks have, during the course of ages, been gradually detached by torrents of water, and brought down in fragments; these, in the violence and rapidity of their descent, are sometimes crumbled to sand, and in this state form the beds of rivers and of the sea, chiefly composed of siliceous materials. Sometimes the fragments are broken without being pulverised by their fall, and assume the form of pebbles, which gradually become rounded and polished.

EMILY.

Pray what is the true colour of silex, which forms such a variety of different coloured substances? Sand is brown, flint is nearly black, and precious stones are of all colours.

MRS. B.

Pure silex, such as is found only in the chemist's laboratory, is perfectly white, and the various colours which it assumes, in the different substances you have just mentioned, proceed from the different ingredients with which it is mixed in them.

CAROLINE.

I wonder that silex is not more valuable, since it forms the basis of so many precious stones.

MRS. B.

You must not forget that the value we set upon precious stones depends in a great measure upon the scarcity with which nature affords them; for, were those productions either common or perfectly imitable by art, they would no longer, notwithstanding their beauty, be so highly esteemed. But the real value of siliceous earth, in many of the most useful arts, is very extensive. Mixed with clay, it forms the basis of all the various kinds of earthen ware, from the most common utensils to the most refined ornaments.

EMILY.

And we must recollect its importance in the formation of glass with potash.

MRS. B.

Nor should we omit to mention, likewise, many other important uses of silex, such as being the chief ingredient of some of the most durable cements, of mortar, &c.

I said before, that siliceous earth combined with no acid but the fluoric; it is for this reason that glass is liable to be attacked by that acid only, which, from its strong affinity for silex, forces that substance from its combination with the potash, and thus destroys the glass.

We will now hasten to proceed to the other earths, for I am rather apprehensive of your growing weary of this part of our subject.

CAROLINE.

The history of the earths is not quite so entertaining as that of the simple substances.

MRS. B.

Perhaps not; but it is absolutely indispensable that you should know something of them; for they form the basis of so many interesing and important compounds, that their total omission would throw great obscurity on our general outline of chemical science. We shall, however, review them in as cursory a manner as the subject can admit of.

ALUMINE derives its name from a compound salt called *alum*, of which it forms the basis.

CAROLINE.

But it ought to be just the contrary, Mrs. B.;

the simple body should give, instead of taking, its name from the compound.

MRS. B.

That is true; but as the compound salt was known long before its basis was discovered, it was very natural that when the earth was at length separated from the acid, it should derive its name from the compound from which it was obtained. However, to remove your scruples, we will call the salt according to the new nomenclature, *sulphat of alumine*. From this combination, alumine may be obtained in its pure state; it is then soft to the touch, makes a paste with water, and hardens in the fire. In nature, it is found chiefly in clay, which contains a considerable proportion of this earth; it is very abundant in fuller's earth, slate, and a variety of other mineral productions. There is indeed scarcely any mineral substance more useful to mankind than alumine. In the state of clay, it forms large strata of the earth, gives consistency to the soil of valleys, and of all low and damp spots, such as swamps and marshes. The beds of lakes, ponds, and springs, are almost entirely of clay; instead of allowing of the filtration of water, as sand does, it forms an impenetrable bottom, and by this means water is accumulated in the caverns of the earth, producing those reservoirs whence springs issue, and spout out at the surface.

EMILY.

I always thought that these subterraneous reservoirs of water were bedded by some hard stone, or rock, which the water could not penetrate.

MRS. B.

That is not the case; for in the course of time water would penetrate, or wear away silex, or any other kind of stone, while it is effectually stopped by clay, or alumine.

The solid compact soils, such as are fit for corn, owe their consistence in a great measure to alumine; this earth is therefore used to improve sandy or chalky soils, which do not retain a sufficient quantity of water for the purpose of vegetation.

Alumine is the most essential ingredient in all potteries. It enters into the composition of brick, as well as that of the finest porcelain; the addition of silex and water hardens it, renders it susceptible of a degree of vitrification, and makes it perfectly fit for its various purposes.

CAROLINE.

I can scarcely conceive that brick and china should be made of the same materials.

MRS. B.

Brick consists almost entirely of baked clay; but a certain proportion of silex is essential to the formation of earthen or stone ware. In common potteries sand is used for that purpose; a more pure silex is, I believe, necessary for the composition of porcelain, as well as a finer kind of clay; and these materials are, no doubt, more carefully prepared, and curiously wrought, in the one case than in the other. Porcelain owes its beautiful semi-transparency to a commencement of vitrification.

EMILY.

But the commonest earthen-ware, though not transparent, is covered with a kind of glazing.

MRS. B.

That precaution is equally necessary for use as for beauty, as the ware would be liable to be spoiled and corroded by a variety of substances, if not covered with a coating of this kind. In porcelain it consists of enamel, which is a fine white opake glass, formed of metallic oxyds, sand, salts, and such other materials as are susceptible of vitrification. The glazing of common earthen-ware is made chiefly of oxyd of lead, or sometimes merely of salt, which, when thinly spread over earthen vessels, will, at a certain heat, run into opake glass.

And of what nature are the colours which are used for painting porcelain.

MRS. B.

They are all composed of metallic oxyds, so that these colours, instead of receiving injury from the application of fire, are strengthened and developed by its action, which causes them to undergo different degrees of oxydation.

Alumine and silex are not only often combined by art, but they have in nature a very strong tendency to unite, and are found combined, in different proportions, in various gems and other minerals. Indeed, many of the precious stones, such as ruby, oriental sapphire, amethyst, &c. consist chiefly of alumine.

We may now proceed to the alkaline earths. I shall say but a few words on BARYTES, as it is hardly ever used, except in chemical laboratories. It is remarkable for its great weight, and its strong alkaline properties, such as destroying animal substances, turning green some blue vegetable colours, and showing a powerful attraction for acids; this last property it possesses to such a degree, particularly with regard to the sulphuric acid, that it will always detect its presence in any substance or combination whatever, by immediately uniting with it, and forming a sulphat of barytes. This renders it a very valuable chemical test. It is found pretty abundantly in nature in the state of carbonat, from which the pure earth can be easily separated.

The next earth we have to consider is LIME. This is a substance of too great and general importance to be passed over so slightly as the last.

Lime is strongly alkaline. In nature it is not met with in its simple state, as its affinity for water and carbonic acid is so great, that it is always found combined with these substances, with which it forms the common lime-stone; but it is separated in the kiln from these ingredients, which are volatilised whenever a sufficient degree of heat is applied.

EMILY.

Pure lime, then, is nothing but lime-stone, which has been deprived, in the kiln, of its water and carbonic acid?

MRS. B.

Precisely: in this state it is called *quick-lime*, and it is so caustic, that it is capable of decomposing the dead bodies of animals very rapidly, without their undergoing the process of putrefaction. — I have here some quick lime, which is kept carefully corked up in a bottle to prevent the access of air; for were it at all exposed to the atmosphere, it would absorb both moisture and carbonic acid gas from it, and be soon slaked. Here is also some lime-stone — we shall pour a little water on each, and observe the effects that result from it.

CAROLINE.

How the quick-lime hisses! It is become excessively hot! — It swells, and now it bursts and crumbles to powder, while the water appears to produce no kind of alteration on the lime-stone.

MRS. B.

Because the lime-stone is already saturated with water, whilst the quick-lime, which has been deprived of it in the kiln, combines with it with very great avidity, and produces this prodigious disengagement of heat, the cause of which I formerly explained to you; do you recollect it?

EMILY.

Yes; you said that the heat did not proceed from the lime, but from the water which was *solidified*, and thus parted with its heat of liquidity.

MRS. B.

Very well. If we continue to add successive quantities of water to the lime after being slaked and crumbled as you see, it will then gradually be diffused in the water, till it will at length be dissolved in it, and entirely disappear; but for this purpose it requires no less than 700 times its weight of water. This solution is called *lime-water*.

CAROLINE.

How very small, then, is the proportion of lime dissolved!

MRS. B.

Barytes is still of more difficult solution; it dissolves only in 900 times its weight of water: but it is much more soluble in the state of crystals. The liquid contained in this bottle is lime-water; it is often used as a medicine, chiefly, I believe, for the purpose of combining with, and neutralising, the superabundant acid which it meets with in the stomach.

EMILY.

I am surprised that it is so perfectly clear; it does not at all partake of the whiteness of the lime.

MRS. B.

Have you forgotten that, in solutions, the solid body is so minutely subdivided by the fluid as to become invisible, and therefore will not in the least degree impair the transparency of the solvent?

I said that the attraction of lime for carbonic acid was so strong, that it would absorb it from the atmosphere. We may see this effect by exposing a glass of lime-water to the air; the lime will then separate from the water, combine with the carbonic acid, and re-appear on the surface in the form of a white film, which is carbonat of lime, commonly called *chalk.*

CAROLINE.

Chalk is, then, a compound salt! I never should have supposed that those immense beds of chalk, that we see in many parts of the country, were a salt. — Now, the white film begins to appear on the surface of the water: but it is far from resembling hard solid chalk.

MRS. B.

That is owing to its state of extreme division; in a little time it will collect into a more compact mass, and subside at the bottom of the glass.

If you breathe into lime-water, the carbonic acid, which is mixed with the air that you expire, will produce the same effect. It is an experiment very easily made; — I shall pour some lime-water into this glass tube, and, by breathing repeatedly into it, you will soon perceive a precipitation of chalk —

EMILY.

I see already a small white cloud formed.

MRS. B.

It is composed of minute particles of chalk; at present it floats in the water, but it will soon subside.

Carbonat of lime, or chalk, you see, is insoluble in water, since the lime which was dissolved re-appears when converted into chalk; but you must take notice of a very singular circumstance, which is, that chalk is soluble in water impregnated with carbonic acid.

CAROLINE.

It is very curious, indeed, that carbonic acid gas should render lime soluble in one instance, and insoluble in the other!

MRS. B.

I have here a bottle of Seltzer water, which, you know, is strongly impregnated with carbonic acid: — let us pour a little of it into a glass of lime-water. You see that it immediately forms a precipitation of carbonat of lime?

EMILY.

Yes, a white cloud appears.

MRS. B.

I shall now pour an additional quantity of the Seltzer water into the lime-water —

EMILY.

How singular! The cloud is re-dissolved, and the liquid is again transparent.

MRS. B.

All the mystery depends upon this circumstance, that carbonat of lime is soluble in carbonic acid, whilst it is insoluble in water; the first quantity of carbonic acid, therefore, which I introduce into the lime-water, was employed in forming the carbonat of lime, which remained visible, until an additional quantity of carbonic acid dissolved it. Thus, you see, when the lime and carbonic acid are in proper proportions to form chalk, the white cloud appears, but when the acid predominates, the chalk is no sooner formed than it is dissolved.

CAROLINE.

That is now the case; but let us try whether a further addition of lime-water will again precipitate the chalk.

EMILY.

It does, indeed! The cloud re-appears, because,

I suppose, there is now no more of the carbonic acid than is necessary to form chalk; and, in order to dissolve the chalk, a superabundance of acid is required.

MRS. B.

We have, I think, carried this experiment far enough; every repetition would but exhibit the same appearances.

Lime combines with most of the acids, to which the carbonic (as being the weakest) readily yields it; but these combinations we shall have an opportunity of noticing more particularly hereafter. It unites with phosphorus, and with sulphur, in their simple state; in short, of all the earths, lime is that which nature employs most frequently, and most abundantly, in its innumerable combinations. It is the basis of all calcareous earths and stones; we find it likewise in the animal and the vegetable creations.

EMILY.

And in the arts is not lime of very great utility?

MRS. B.

Scarcely any substance more so; you know that it is a most essential requisite in building, as it constitutes the basis of all cements, such as mortar, stucco, plaister, &c.

Lime is also of infinite importance in agriculture; it lightens and warms soils that are too cold, and compact, in consequence of too great a proportion of clay. — But it would be endless to enumerate the various purposes for which it is employed; and you know enough of it to form some idea of its importance; we shall, therefore, now proceed to the third alkaline earth, MAGNESIA.

CAROLINE.

I am already pretty well acquainted with that earth; it is a medicine.

MRS. B.

It is in the state of carbonat that magnesia is usually employed medicinally; it then differs but little in appearance from its simple form, which is that of a very fine light white powder. It dissolves in 2000 times its weight of water, but forms with acids extremely soluble salts. It has not so great an attraction for acids as lime, and consequently yields them to the latter. It is found in a great variety of mineral combinations, such as slate, mica, amianthus, and more particularly in a certain lime stone, which has lately been discovered by Mr. Tennant to contain it in very great quantities. It does not attract and solidify water, like lime: but when mixed with water and exposed to the atmosphere, it slowly absorbs carbonic acid from the latter, and thus loses its causticity. Its chief use in medicine is, like that of lime, derived from its readiness to combine with, and neutralise, the acid which it meets with in the stomach.

EMILY.

Yet, you said that it was taken in the state of carbonat, in which case it has already combined with an acid?

MRS. B.

Yes; but the carbonic is the last of all the acids in the order of affinities; it will therefore yield the magnesia to any of the others. It is, however, frequently taken in its caustic state as a remedy for flatulence. Combined with sulphuric acid, magnesia forms another and more powerful medicine, commonly called *Epsom salt.*

CAROLINE.

And properly, *sulphat of magnesia*, I suppose? Pray why was it ever called Epsom salt?

MRS. B.

Because there is a spring in the neighbourhood of Epsom which contains this salt in great abundance.

The last alkaline earth which we have to mention is STRONTIAN, or STRONTITES, discovered by

Dr. Hope a few years ago. It so strongly resembles barytes in its properties, and is so sparingly found in nature, and of so little use in the arts, that it will not be necessary to enter into any particulars respecting it. One of the remarkable characteristic properties of strontites is, that its salts, when dissolved in spirit of wine, tinge the flame of a deep red, or blood colour.

CONVERSATION XVI.

ON ACIDS.

MRS. B.

We may now proceed to the acids. Of the metallic oxyds, you have already acquired some general notions. This subject, though highly interesting in its details, is not of sufficient importance to our concise view of chemistry, to be particularly treated of; but it is absolutely necessary that you should be better acquainted with the acids, and likewise with their combinations with the alkalies, which form the triple compounds called NEUTRAL SALTS.

The class of acids is characterised by very distinct properties. They all change blue vegetable infusions to a red colour: they are all more or less sour to the taste; and have a general tendency to combine with the earths, alkalies, and metallic oxyds.

You have, I believe, a clear idea of the nomenclature by which the base (or radical) of the acid, and the various degrees of acidification, are expressed?

EMILY.

Yes, I think so; the acid is distinguished by the

name of its base, and its degree of oxydation, that is, the quantity of oxygen it contains, by the termination of that name in *ous* or *ic*; thus sulphureous acid is that formed by the smallest proportion of oxygen combined with sulphur; sulphur*ic* acid that which results from the combination of sulphur with the greatest quantity of oxygen.

MRS. B.

A still greater latitude may, in many cases, be allowed to the proportions of oxygen than can be combined with acidificiable radicals; for several of these radicals are susceptible of uniting with a quantity of oxygen so small as to be insufficient to give them the properties of acids; in these cases, therefore, they are converted into oxyds. Such is sulphur, which by exposure to the atmosphere with a degree of heat inadequate to produce inflammation, absorbs a small proportion of oxygen, which colours it red or brown. This, therefore, is the first degree of oxygenation of sulphur; the 2d converts it into sulphur*ous* acid; the 3d into the sulphur*ic* acid; and 4thly, if it was found capable of combining with a still larger proportion of oxygen, it would then be termed *super-oxygenated sulphuric acid.*

EMILY.

Are these various degrees of oxygenation common to all the acids?

MRS. B.

No; they vary much in this respect: some are susceptible of only one degree of oxygenation; others, of two, or three; there are but very few that will admit of more.

CAROLINE.

The modern nomenclature must be of immense advantage in pointing out so easily the nature of the acids, and their various degrees of oxygenation.

MRS. B.

Till lately many of the acids had not been decomposed; but analogy afforded so strong a proof of their compound nature, that I never could reconcile myself to classing them with the simple bodies, though this division has been adopted by several chemical writers. At present there are only the muriatic and the fluoric acids, which have not had their bases distinctly separated.

CAROLINE.

We have heard of a great variety of acids; pray how many are there in all?

MRS. B.

I believe there are reckoned at present thirty-four, and their number is constantly increasing, as the science improves; but the most important, and those to which we shall almost entirely confine our attention, are but few. I shall, however, give you a general view of the whole; and then we shall more particularly examine those that are the most essential.

This class of bodies was formerly divided into mineral, vegetable, and animal acids, according to the substances from which they were commonly obtained.

CAROLINE.

That, I should think, must have been an excellent arrangement; why was it altered?

MRS. B.

Because in many cases it produced confusion. In which class, for instance, would you place carbonic acid?

CAROLINE.

Now I see the difficulty. I should be at a loss where to place it, as you have told us that it exists in the animal, vegetable, and mineral kingdoms.

EMILY.

There would be the same objection with respect to phosphoric acid, which, though obtained chiefly from bones, can also, you said, be found in small quantities in stones, and likewise in some plants.

MRS. B.

You see, therefore, the propriety of changing

this mode of classification, These objections do not exist in the present nomenclature; for the composition and nature of each individual acid is in some degree pointed out, instead of the class of bodies from which it is extracted; and, with regard to the more general division of acids, they are classed under these three heads:

First, Acids of known or supposed simple bases, which are formed by the union of these bases with oxygen. They are the following:

The *Sulphuric* *Carbonic* *Nitric* *Phosphoric* *Arsenical* *Tungstenic* *Molybdenic* *Boracic* *Fluoric* *Muriatic*	Acids, o known and simple bases.

This class comprehends the most anciently known and most important acids. The sulphuric, nitric, and muriatic were formerly, and are still frequently, called *mineral acids.*

2dly, Acids that have double or binary radicals, and which consequently consist of triple combinations. These are the vegetable acids, whose common radical is a compound of hydrogen and carbon.

CAROLINE.

But if the basis of all the vegetable acids be the same, it should form but one acid; it may indeed combine with different proportions of oxygen, but the nature of the acid must be the same.

MRS. B.

The only difference that exists in the basis of vegetable acids, is the various proportions of hydrogen and carbon from which they are severally composed. But this is enough to produce a number of acids apparently very dissimilar. That they do not, however, differ essentially, is proved by their susceptibility of being converted into each other, by the addition or subtraction of a portion of hydrogen or of carbon. The names of these acids are,

The *Acetic* *Oxalic* *Tartarous* *Citric* *Malic* *Gallic* *Mucous* *Benzoic* *Succinic* *Camphoric* *Suberic*	Acids, of double bases, being of vegetable origin.

The 3d class of acids consists of those which have triple radicals, and are therefore of a still more compound nature. This class comprehends the animal acids, which are,

The *Lactic* *Prussic* *Formic* *Bombic* *Sebacic* *Zoonic* *Lithic*	Acids, of triple bases, or animal acids.

I have given you this summary account or enumeration of the acids, as you may find it more satisfactory to have at once an outline or a general notion of the extent of the subject; but we shall now confine ourselves to the first class, which requires our more immediate attention; and defer the few remarks which we shall have to make on the others, till we treat of the chemistry of the animal and vegetable kingdoms.

The acids of simple and known radicals are all capable of being decomposed by combustible bodies, to which they yield their oxygen. If, for instance, I pour a drop of sulphuric acid on this piece of iron, it will produce a spot of rust, you know what that is?

CAROLINE.

Yes; it is an oxyd, formed by the oxygen of the acid combining with the iron.

MRS. B.

In this case you see the sulphur deposits the oxygen by which it was acidified on the metal. And again, if we pour some acid on a compound combustible substance, (we shall try it on this piece of wood,) it will combine with one or more of the constituents of that substance, and occasion a decomposition.

EMILY.

It has changed the colour of the wood to black. How is that?

MRS. B.

The oxygen deposited by the acid has burnt it; you know that wood in burning becomes black before it is reduced to ashes. Whether it derives the oxygen which burns it from the atmosphere, or from any other source, the chemical effect on the wood is the same. In the case of real combustion, wood becomes black, because it is reduced to the state of charcoal by the evaporation of its other constituents. But can you tell me the reason why wood turns black when burnt by the application of an acid?

CAROLINE.

First, tell me what are the ingredients of wood?

MRS. B.

Hydrogen and carbon are the chief constituents of wood, as of all other vegetable substances.

CAROLINE.

Well, then, I suppose that the oxygen of the acid combines with the hydrogen of the wood, to form water; and that the carbon of the wood, remaining alone, appears of its usual black colour.

MRS. B.

Very well indeed, my dear; that is certainly the most plausible explanation.

EMILY.

Would not this be a good method of making charcoal?

MRS. B.

It would be an extremely expensive, and, I believe, very imperfect method; for the action of the acid on the wood, and the heat produced by it, are far from sufficient to deprive the wood of all its evaporable parts.

CAROLINE.

What is the reason that vinegar, lemon, and the acid of fruits, do not produce this effect on wood?

MRS. B.

They are vegetable acids, whose bases are composed of hydrogen and carbon; the oxygen, therefore, will not be disposed to quit this radical, where it is already united with hydrogen. The strongest of these may, perhaps, yield a little of their oxygen to the wood, and produce a stain upon it; but the carbon will not be sufficiently uncovered to assume its black colour. Indeed, the several mineral acids themselves possess this power of charing wood in very different degrees.

EMILY.

Cannot vegetable acids be decomposed, by any combustibles?

MRS. B.

No; because their radical is composed of two substances which have a greater attraction for oxy gen than any known body.

CAROLINE.

And are those strong acids, which burn and decompose wood, capable of producing similar effects on the skin and flesh of animals?

MRS. B.

Yes; all the mineral acids, and one of them more especially, possess powerful caustic qualities. They actually corrode and destroy the skin and flesh; but they do not produce upon these exactly the same alteration they do on wood, probably because there is a great proportion of nitrogen and other substances in animal matter, which prevents the separation of carbon from being so conspicuous.

CONVERSATION XVII.

OF THE SULPHURIC AND PHOSPHORIC ACIDS; OR THE COMBINATIONS OF OXYGEN WITH SULPHUR AND PHOSPHORUS; AND OF THE SULPHATS AND PHOSPHATS.

MRS. B.

In addition to the general survey which we have taken of acids, I think you will find it interesting to examine individually a few of the most important of them, and likewise some of their principal combinations with the alkalies, alkaline earths, and metals. The first of the acids, in point of importance, is the SULPHURIC, formerly called *oil of vitriol.*

CAROLINE.

I have known it a long time by that name, but had no idea that it was the same fluid as sulphuric acid. What resemblance or connection can there be between oil of vitriol and this acid?

MRS. B.

Vitriol is the common name for sulphat of iron, a salt which is formed by the combination of sulphuric acid and iron; the sulphuric acid was formerly obtained by distillation from this salt, and it very naturally received its name from the substance which afforded it.

CAROLINE.

But it is still usually called oil of vitriol?

MRS. B.

Yes; a sufficient length of time has not yet elapsed, since the invention of the new nomenclature, for it to be generally disseminated; but, as it is adopted by all scientific chemists, there is every reason to suppose that it will gradually become universal. When I received this bottle from the chemists, *oil of vitriol* was inscribed on the label; but, as I knew you were very punctilious in regard to the nomenclature, I changed it, and substituted the words *sulphuric acid.*

EMILY.

This acid has neither colour nor smell, but it appears much thicker than water.

MRS. B.

It is nearly twice as heavy as water, and has, you see, an oily consistence.

CAROLINE.

And it is probably from this circumstance that

it has been called an oil, for it can have no real claim to that name, as it does not contain either hydrogen or carbon, which are the essential constituents of oil.

MRS. B.

Certainly; and therefore it would be the more absurd to retain a name which owed its origin to such a mistaken analogy.

Sulphuric acid, in its purest state, would probably be a concrete substance, but its attraction for water is such, that it is impossible to obtain that acid perfectly free from it; it is, therefore, always seen in a liquid form, such as you here find it. One of the most striking properties of sulphuric acid is that of evolving a considerable quantity of heat when mixed with water; this I have already shown you.

EMILY.

Yes, I recollect it; but what was the degree of heat produced by that mixture?

MRS. B.

The thermometer may be raised by it to 300 degrees, which is considerably above the temperature of boiling water.

CAROLINE.

Then water might be made to boil in that mixture?

MRS. B.

Nothing more easy, provided that you employ sufficient quantities of acid and of water, and in the due proportions. The greatest heat is produced by a mixture of one part of water to four of the acid: we shall make a mixture of these proportions, and immerse in it this thin glass tube, which is full of water.

CAROLINE.

The vessel feels extremely hot, but the water does not boil yet.

MRS. B.

You must allow some time for the heat to penetrate the tube, and raise the temperature of the water to the boiling point —

CAROLINE.

Now it boils — and with increasing violence.

MRS. B.

But it will not continue boiling long; for the mixture gives out heat only while the particles of the water and the acid are mutually penetrating each other: as soon as the new arrangement of those particles is effected, the mixture will gradually cool, and the water return to its former temperature.

You have seen the manner in which sulphuric

acid decomposes all combustible substances, whether animal, vegetable, or mineral, and burns them by means of its oxygen?

CAROLINE.

I have very unintentionally repeated the experiment on my gown, by letting a drop of the acid fall upon it, and it has made a stain, which, I suppose, will never wash out.

MRS. B.

No, certainly; for before you can put it into water, the spot will become a hole, as the acid has literally burnt the muslin.

CAROLINE.

So it has, indeed! Well, I will fasten the stopper, and put the bottle away, for it is a dangerous substance. — Oh, now I have done worse still, for I have spilt some on my hand!

MRS. B.

It is then burned, as well as your gown, for you know that oxygen destroys animal as well as vegetable matters; and, as far as the decomposition of the skin of your finger is effected, there is no remedy; but by washing it immediately in water, you will dilute the acid, and prevent any further injury.

CAROLINE.

It feels extremely hot, I assure you.

MRS. B.

You have now learned, by experience, how cautiously this acid must be used. You will soon become acquainted with another acid, the nitric, which, though it produces less heat on the skin, destroys it still quicker, and makes upon it an indelible stain. You should never handle any substances of this kind, without previously dipping your fingers in water, which will weaken their caustic effects. But, since you will not repeat the experiment, I must put in the stopper, for the acid attracts the moisture from the atmosphere, which would destroy its strength and purity.

EMILY.

Pray, how can sulphuric acid be extracted from sulphat of iron by distillation?

MRS. B.

The process of distillation, you know, consists in separating substances from one another by means of their different degrees of volatility, and by the introduction of a new chemical agent, caloric. Thus, if sulphat of iron be exposed in a retort to a proper degree of heat, it will be decomposed, and the sulphuric acid will be volatilised.

EMILY.

But now that the process of forming acids by the combustion of their radicals is known, why should not this method be used for making sulphuric acid?

MRS. B.

This is actually done in most manufactures; but the usual method of preparing sulphuric acid does not consist in burning the sulphur in oxygen gas (as we formerly did by the way of experiment), but in heating it together with another substance, nitre, which yields oxygen in sufficient abundance to render the combustion in common air rapid and complete.

CAROLINE.

This substance, then, answers the same purpose as oxygen gas?

MRS. B.

Exactly. In manufactures the combustion is performed in a leaden chamber, with water at the bottom, to receive the vapour and assist its condensation. The combustion is, however, never so perfect but that a quantity of *sulphureous* acid is formed at the same time; for you recollect that the sulphureous acid, according to the chemical nomenclature, differs from the sulphuric only by containing less oxygen.

From its own powerful properties, and from the various combinations into which it enters, sulphuric acid is of great importance in many of the arts.

It is used also in medicine in a state of great dilution; for were it taken internally, in a concentrated state, it would prove a most dangerous poison.

CAROLINE.

I am sure it would burn the throat and stomach.

MRS. B.

Can you think of any thing that would prove an antidote to this poison?

CAROLINE.

A large draught of water to dilute it.

MRS. B.

That would certainly weaken the caustic power of the acid, but it would increase the heat to an intolerable degree. Do you recollect nothing that would destroy its deleterious properties more effectually?

EMILY.

An alkali might, by combining with it; but, then, a pure alkali is itself a poison, on account of its causticity.

MRS. B.

There is no necessity that the alkali should be

caustic. Soap, in which it is combined with oil; or magnesia, either in the state of carbonat, or mixed with water, would prove the best antidotes.

EMILY.

In those cases then, I suppose, the potash and the magnesia would quit their combinations to form salts with the sulphuric acid?

MRS. B.

Precisely.

We may now make a few observations on the sulphure*ous* acid, which we have found to be the product of sulphur slowly and imperfectly burnt. This acid is distinguished by its pungent smell, and its gaseous form.

CAROLINE.

Its aëriform state is, I suppose, owing to the smaller proportion of oxygen, which renders it lighter than sulphur*ic* acid?

MRS. B.

Probably; for by adding oxygen to the weaker acid; it may be converted into the stronger kind. But this change of state may also be connected with a change of affinity with regard to caloric.

EMILY.

And may sulphureous acid be obtained from sulphuric acid by a diminution of oxygen?

MRS. B.

Yes; it can be done by bringing any combustible substance in contact with the acid. This decomposition is most easily performed by some of the metals; these absorb a portion of the oxygen from the sulphuric acid, which is thus converted into the sulphureous, and flies off in its gaseous form.

CAROLINE.

And cannot the sulphureous acid itself be decomposed and reduced to sulphur?

MRS. B.

Yes; if this gas be heated in contact with charcoal, the oxygen of the gas will combine with it, and the pure sulphur is regenerated.

Sulphureous acid is readily absorbed by water; and in this liquid state it is found particularly useful in bleaching linen and woollen cloths, and is much used in manufactures for those purposes. I can show you its effect in destroying colours, by taking out vegetable stains — I think I see a spot on your gown, Emily, on which we may try the experiment.

EMILY.

It is the stain of mulberries; but I shall be almost afraid of exposing my gown to the experiment, after seeing the effect which the sulphuric acid produced on that of Caroline —

MRS. B.

There is no such danger from the sulphureous; but the experiment must be made with great caution, for, during the formation of sulphureous acid by combustion, there is always some sulphuric produced.

CAROLINE.

But where is your sulphureous acid?

MRS. B.

We may easily prepare some ourselves, simply by burning a match; we must first wet the stain with water, and now hold it in this way, at a little distance, over the lighted match: the vapour that arises from it is sulphureous acid, and the stain, you see, gradually disappears.

EMILY.

I have frequently taken out stains by this means, without understanding the nature of the process. But why is it necessary to wet the stain before it is exposed to the acid fumes?

MRS. B.

The moisture attracts and absorbs the sulphureous acid; and it serves likewise to dilute any particles of sulphuric acid which might injure the linen.

Sulphur is susceptible of a third combination

with oxygen, in which the proportion of the latter is too small to render the sulphur acid. It acquires this slight oxygenation by mere exposure to the atmosphere, without any elevation of temperature: in this case, the sulphur does not change its natural form, but is only discoloured, being changed to red or brown; and in this state it is an oxyd of sulphur.

Before we take leave of the sulphuric acid, we shall say a few words of its principal combinations. It unites with all the alkalies, alkaline earths and metals, to form compound salts.

CAROLINE.

Pray, give me leave to interrupt you for a moment: you have never mentioned any other salts than the compound or neutral salts; is there no other kind?

MRS. B.

The term *salt* has been used, from time immemorial, as a kind of general name for any substance that has savour, odour, is soluble in water, and crystallisable, whether it be of an acid, an alkaline, or compound nature; but the compound salts alone retain that appellation in modern chemistry.

The most important of the salts, formed by the combinations of the sulphuric acid, are, first, *sulphat of potash*, formerly called *sal polychrest*: this is a very bitter salt, much used in medicine; it is found in the ashes of most vegetables, but it may be prepared artificially by the immediate combination of sulphuric acid and potash. This salt is easily soluble in boiling water. Solubility is, indeed, a property common to all salts; and they always produce cold in melting.

EMILY.

That must be owing to the caloric which they absorb in passing from a solid to a fluid form.

MRS. B.

That is, certainly, the most probable explanation.

Sulphat of soda, commonly called Glauber's salt, is another medicinal salt, which is still more bitter than the preceding. We must prepare some of these compounds, that you may observe the phenomena which take place during their formation. We need only pour some sulphuric acid over the soda whieh I have put into this glass.

CAROLINE.

What an amazing heat is disengaged! — I thought you said that cold was produced by the melting of salts?

MRS. B.

But you must observe that we are now *making*,

not *melting* a salt. Heat is disengaged during the formation of compound salts, and a faint light is also emitted, which may sometimes be perceived in the dark.

EMILY.

And is this heat and light produced by the union of the opposite electricities of the alkali and the acid?

MRS. B.

No doubt it is, if that theory be true.

CAROLINE.

The union of an acid and an alkali is then an actual combustion?

MRS. B.

Not precisely, though there is certainly much analogy in these processes.

CAROLINE.

Will this sulphat of soda become solid?

MRS. B.

We have not, I suppose, mixed the acid and the alkali in the exact proportions that are required for the formation of the salt, otherwise the mixture would have been almost immediately changed to a solid mass; but, in order to obtain it in crystals, as you see it in this bottle, it would be necessary first to dilute it with water, and afterwards to evaporate the water, during which operation the salt would gradually crystallise.

CAROLINE.

But of what use is the addition of water, if it is afterwards to be evaporated?

MRS. B.

When suspended in water, the acid and the alkali are more at liberty to act on each other, their union is more complete, and the salt assumes the regular form of crystals during the slow evaporation of its solvent.

Sulphat of soda liquefies by heat, and effloresces in the air.

EMILY.

Pray what is the meaning of the word *effloresces?* I do not recollect your having mentioned it before.

MRS. B.

A salt is said to effloresce when it loses its water of crystallisation on being exposed to the atmosphere, and is thus gradually converted into a dry powder: you may observe that these crystals of sulphat of soda are far from possessing the transparency which belongs to their crystalline state; they are covered with a white powder, occasioned by their having been exposed to the atmosphere, which has deprived their surface of its lustre, by absorbing its water of crystallisation. Salts are, in general, either *efflorescent* or *deliquescent*: this latter property is precisely the reverse of the former; that is to say, deliquescent salts absorb water from the atmosphere, and are moistened and gradually melted by it. Muriat of lime is an instance of great deliquescence.

EMILY.

But are there no salts that have the same degree of attraction for water as the atmosphere, and that will consequently not be affected by it?

MRS. B.

Yes; there are many such salts, as, for instance, common salt, sulphat of magnesia, and a variety of others.

Sulphat of lime is very frequently met with in nature, and constitutes the well-known substance called *gypsum*, or *plaster of Paris*.

Sulphat of magnesia, commonly called *Epsom salt*, is another very bitter medicine, which is obtained from sea-water and from several springs, or may be prepared by the direct combination of its ingredients.

We have formerly mentioned *sulphat of alumine* as constituting the common *alum*; it is found in nature chiefly in the neighbourhood of volcanos, and is particularly useful in the arts, from its strong astringent qualities. It is chiefly employed by dyers and calico-printers, to fix colours; and is used also in the manufacture of some kinds of leather.

Sulphuric acid combines also with the metals.

CAROLINE.

One of these combinations, *sulphat of iron*, we are already well acquainted with.

MRS. B.

That is the most important metallic salt formed by sulphuric acid, and the only one that we shall here notice. It is of great use in the arts; and, in medicine, it affords a very valuable tonic: it is of this salt that most of those preparations called *steel medicines* are composed.

CAROLINE.

But does any carbon enter into these compositions to form steel?

MRS. B.

Not an atom: they are, therefore, very improperly called steel: but it is the vulgar appellation, and medical men themselves often comply with the general custom.

Sulphat of iron may be prepared, as you have seen, by dissolving iron in sulphuric acid; but it is generally obtained from the natural production called *Pyrites*, which being a sulphuret of iron, requires only exposure to the atmosphere to be oxydated, in order to form the salt; this, therefore, is much the most easy way of procuring it on a large scale.

EMILY.

I am surprised to find that both acids and compound salts are generally obtained from their various combinations, rather than from the immediate union of their ingredients.

MRS. B.

Were the simple bodies always at hand, their combinations would naturally be the most convenient method of forming compounds; but you must consider that, in most instances, there is great difficulty and expense in obtaining the simple ingredients from their combinations; it is, therefore, often more expedient to procure compounds from the decomposition of other compounds. But, to return to the sulphat of iron. — There is a certain vegetable acid called *Gallic acid*, which has the remarkable property of precipitating this salt black — I shall pour a few drops of the gallic acid into this solution of sulphat of iron —

CAROLINE.

It is become as black as ink!

MRS. B.

And it is ink in reality. Common writing ink is a precipitate of sulphat of iron by gallic acid; the black colour is owing to the formation of gallat of iron, which being insoluble, remains suspended in the fluid.

This acid has also the property of altering the colour of iron in its metallic state. You may frequently see its effect on the blade of a knife, that has been used to cut certain kinds of fruits.

CAROLINE.

True; and that is, perhaps, the reason that a silver knife is preferred to cut fruits; the gallic acid, I suppose, does not act upon silver. — Is this acid found in all fruits?

MRS. B.

It is contained, more or less, in the rind of most fruits and roots, especially the radish, which, if scraped with a steel or iron knife, has its bright red colour changed to a deep purple, the knife being at the same time blackened. But the vegetable substance in which the gallic acid most abounds is *nutgall*, a kind of excrescence that grows on oaks, and from which the acid is commonly obtained for its various purposes.

MRS. B.

We now come to the PHOSPHORIC and PHOSPHOROUS ACIDS. In treating of phosphorus, you have seen how these acids may be obtained from it by combustion?

EMILY.

Yes; but I should be much surprised if it was the usual method of obtaining them, since it is so very difficult to procure phosphorus in its pure state.

MRS. B.

You are right, my dear; the phosphoric acid, for general purposes, is extracted from bones, in which it is contained in the state of phosphat of lime; from this salt the phosphoric acid is separated by means of the sulphuric, which combines with the lime. In its pure state, phosphoric acid is either liquid or solid, according to its degree of concentration.

Among the salts formed by this acid, *phosphat of lime* is the only one that affords much interest; and this, we have already observed, constitutes the basis of all bones. It is also found in very small quantities in some vegetables.

CONVERSATION XVIII.

OF THE NITRIC AND CARBONIC ACIDS: OR THE COMBINATIONS OF OXYGEN WITH NITROGEN AND CARBON; AND OF THE NITRATS AND CARBONATS.

MRS. B.

I AM almost afraid of introducing the subject of the NITRIC ACID, as I am sure that I shall be blamed by Caroline for not having made her acquainted with it before.

CAROLINE.

Why so, Mrs. B.?

MRS. B.

Because you have long known its radical, which is nitrogen or azote; and in treating of that element, I did not even hint that it was the basis of an acid.

CAROLINE.

And what could be your reason for not mention- this acid sooner?

MRS. B.

I do not know whether you will think the reason

sufficiently good to acquit me; but the omission, I assure you, did not proceed from negligence. You may recollect that nitrogen was one of the first simple bodies which we examined; you were then ignorant of the theory of combustion, which I believe was, for the first time, mentioned in that lesson; and therefore it would have been in vain, at that time, to have attempted to explain the nature and formation of acids.

CAROLINE.

I wonder, however, that it never occurred to us to enquire whether nitrogen could be acidified; for, as we knew it was classed among the combustible bodies, it was natural to suppose that it might produce an acid.

MRS. B.

That is not a necessary consequence; for it might combine with oxygen only in the degree requisite to form an oxyd. But you will find that nitrogen is susceptible of various degrees of oxygenation, some of which convert it merely into an oxyd, and others give it all the acid properties.

The acids, resulting from the combination of oxygen and nitrogen, are called the NITROUS and NITRIC acids. We will begin with the NITRIC, in which nitrogen is in the highest state of oxygenation. This acid naturally exists in the form of gas; but is so very soluble in water, and has so great an affinity for it, that one grain of water will absorb and condense ten grains of acid gas, and form the limpid fluid which you see in this bottle.

CAROLINE.

What a strong offensive smell it has!

MRS. B.

This acid contains a greater abundance of oxygen than any other, but it retains it with very little force.

EMILY.

Then it must be a powerful caustic, both from the facility with which it parts with its oxygen, and the quantity which it affords?

MRS. B.

Very well, Emily; both cause and effect are exactly such as you describe: nitric acid burns and destroys all kinds of organised matter. It even sets fire to some of the most combustible substances.—We shall pour a little of it over this piece of dry warm charcoal—you see it inflames it immediately; it would do the same with oil of turpentine, phosphorus, and several other very combustible bodies. This shows you how easily this acid is decomposed by combustible bodies, since these effects must depend upon the absorption of its oxygen.

Nitric acid has been used in the arts from time immemorial, but it is only within these twenty-five years that its chemical nature has been ascertained. The celebrated Mr. Cavendish discovered that it consisted of about 10 parts of nitrogen and 25 of oxygen. * These principles, in their gaseous state, combine at a high temperature; and this may be effected by repeatedly passing the electrical spark through a mixture of the two gases.

EMILY.

The nitrogen and oxygen gases, of which the atmosphere is composed, do not combine, I suppose, because their temperature is not sufficiently elevated?

CAROLINE.

But in a thunder-storm, when the lightning repeatedly passes through them, may it not produce nitric acid? We should be in a strange situation, if a violent storm should at once convert the atmosphere into nitric acid.

MRS. B.

There is no danger of it, my dear; the lightning can affect but a very small portion of the at-

* The proportion stated by Sir H. Davy, in his Chemical Researches, is as 1 to 2,389.

mosphere, and though it were occasionally to produce a little nitric acid, yet this never could happen to such an extent as to be perceivable.

EMILY.

But how could the nitric acid be known, and used, before the method of combining its constituents was discovered?

MRS. B.

Before that period the nitric acid was obtained, and it is indeed still extracted, for the common purposes of art, from the compound salt which it forms with potash, commonly called *nitre.*

CAROLINE.

Why is it so called? Pray, Mrs. B., let these old unmeaning names be entirely given up, by us at least; and let us call this salt *nitrat of potash.*

MRS. B.

With all my heart; but it is necessary that I should, at least, mention the old names, and more especially those which are yet in common use; otherwise, when you meet with them, you would not be able to understand their meaning.

EMILY.

And how is the acid obtained from this salt?

MRS. B.

By the intervention of sulphuric acid, which combines with the potash, and sets the nitric acid at liberty. This I can easily show you, by mixing some nitrat of potash and sulphuric acid in this retort, and heating it over a lamp; the nitric acid will come over in the form of vapour, which we shall collect in a glass bell. This acid, diluted in water, is commonly called *aqua fortis*, if Caroline will allow me to mention that name.

CAROLINE.

I have often heard that aqua fortis will dissolve almost all metals; it is no doubt because it yields its oxygen so easily.

MRS. B.

Yes; and from this powerful solvent property, it derived the name of aqua fortis, or strong water. Do you not recollect that we oxydated, and afterwards dissolved, some copper in this acid?

EMILY.

If I remember right, the nitrat of copper was the first instance you gave us of a compound salt.

CAROLINE.

Can the nitric acid be completely decomposed and converted into nitrogen and oxygen?

EMILY.

That cannot be the case, Caroline; since the acid can be decomposed only by the combination of its constituents with other bodies.

MRS. B.

True; but caloric is sufficient for this purpose. By making the acid pass through a red hot porcelain tube, it is decomposed; the nitrogen and oxygen regain the caloric which they had lost in combining, and are thus both restored to their gaseous state.

The nitric acid may also be partly decomposed, and is by this means converted into NITROUS ACID.

CAROLINE.

This conversion must be easily effected, as the oxygen is so slightly combined with the nitrogen.

MRS. B.

The partial decomposition of nitric acid is readily effected by most metals; but it is sufficient to expose the nitric acid to a very strong light to make it give out oxygen gas, and thus be converted into nitrous acid. Of this acid there are various degrees, according to the proportions of oxygen which it contains; the strongest, and that into which the nitric is first converted, is of a yellow colour, as you see in this bottle.

CAROLINE.

How it fumes when the stopper is taken out!

MRS. B.

The acid exists naturally in a gaseous state, and is here so strongly concentrated in water, that it is constantly escaping.

Here is another bottle of nitrous acid, which, you see, is of an orange red; this acid is weaker, the nitrogen being combined with a smaller quantity of oxygen; and with a still less proportion of oxygen it is an olive-green colour, as it appears in this third bottle. In short, the weaker the acid, the deeper is its colour.

Nitrous acid acts still more powerfully on some inflammable substances than the nitric.

EMILY.

I am surprised at that, as it contains less oxygen.

MRS. B.

But, on the other hand, it parts with its oxygen much more readily: you may recollect that we once inflamed oil with this acid.

The next combinations of nitrogen and oxygen form only oxyds of nitrogen, the first of which is commonly called *nitrous air*; or more properly *nitric oxyd gas*. This may be obtained from nitric acid, by exposing the latter to the action of metals, as in dissolving them it does not yield the whole of its oxygen, but retains a portion of this principle sufficient to convert it into this peculiar gas, a specimen of which I have prepared, and preserved within this inverted glass bell.

EMILY.

It is a perfectly invisible elastic fluid.

MRS. B.

Yes; and it may be kept any length of time in this manner over water, as it is not, like the nitric and nitrous acids, absorbable by it. It is rather heavier than atmospherical air, and is incapable of supporting either combustion or respiration. I am going to incline the glass gently on one side, so as to let some of the gas escape —

EMILY.

How very curious! — It produces orange fumes like the nitrous acid! that is the more extraordinary, as the gas within the glass is perfectly invisible.

MRS. B.

It would give me much pleasure if you could make out the reason of this curious change without requiring any further explanation.

CAROLINE.

It seems, by the colour and smell, as if it were

converted into nitrous acid gas: yet that cannot be, unless it combines with more oxygen; and how can it obtain oxygen the very instant it escapes from the glass?

EMILY.

From the atmosphere, no doubt. Is it not so, Mrs. B.?

MRS. B.

You have guessed it; as soon as it comes in contact with the atmosphere, it absorbs from it the additional quantity of oxygen necessary to convert it into nitrous acid gas. And, if I now remove the bottle entirely from the water, so as to bring at once the whole of the gas into contact with the atmosphere, this conversion will appear still more striking —

EMILY.

Look, Caroline, the whole capacity of the bottle is instantly tinged of an orange colour!

MRS. B.

Thus, you see, it is the most easy process imaginable to convert *nitrous oxyd gas* into *nitrous acid gas.* The property of attracting oxygen from the atmosphere, without any elevation of temperature, has occasioned this gaseous oxyd being used as a test for ascertaining the degree of purity of the atmosphere. I am going to show you how it is applied to this purpose.—You see this graduated glass tube, which is closed at one end, (PLATE X. Fig. 2.)—I first fill it with water, and then introduce a certain measure of nitrous gas, which, not being absorbable by water, passes through it, and occupies the upper part of the tube. I must now add rather above two-thirds of oxygen gas, which will just be sufficient to convert the nitrous oxyd gas into nitrous acid gas.

CAROLINE.

So it has! — I saw it turn of an orange colour; but it immediately afterwards disappeared entirely, and the water, you see, has risen, and almost filled the tube.

MRS. B.

That is because the acid gas is absorbable by water, and in proportion as the gas impregnates the water, the latter rises in the tube. When the oxygen gas is very pure, and the required proportion of nitrous oxyd gas very exact, the whole is absorbed by the water; but if any other gas be mixed with the oxygen, instead of combining with the nitrous oxygen, it will remain and occupy the upper part of the tube; or, if the gases be not in the due proportion, there will be a residue of that which predominates. — Before we leave this subject, I must not forget to remark that nitrous acid may be formed by dissolving nitrous oxyd gas in nitric acid. This solution may be effected simply by making bubbles of nitrous oxyd gas pass through nitric acid.

EMILY.

That is to say, that nitrogen at its highest degree of oxygenation, being mixed with nitrogen at its lowest degree of oxygenation, will produce a kind of intermediate substance, which is nitrous acid.

MRS. B.

You have stated the fact with great precision. — There are various other methods of preparing nitrous oxyd, and of obtaining it from compound bodies; but it is not necessary to enter into these particulars. It remains for me only to mention another curious modification of oxygenated nitrogen, which has been distinguished by the name of *gaseous oxyd of nitrogen.* It is but lately that this gas has been accurately examined, and its properties have been investigated chiefly by Sir H. Davy. It has obtained also the name of *exhilarating* gas, from the very singular property which that gentleman has discovered in it, of elevating the animal spirits, when inhaled into the lungs, to a degree sometimes resembling delirium or intoxication.

CAROLINE.

Is it respirable, then?

MRS. B.

It can scarcely be called respirable, as it would not support life for any length of time; but it may be breathed for a few moments without any other effects, than the singular exhilaration of spirits I have just mentioned. It affects different people, however, in a very different manner. Some become violent, even outrageous: others experience a languor, attended with faintness; but most agree in opinion, that the sensations it excites are extremely pleasant.

CAROLINE.

I think I should like to try it — how do you breathe it?

MRS. B.

By collecting the gas in a bladder, to which a short tube with a stop-cock is adapted; this is applied to the mouth with one hand, whilst the nostrils are kept closed with the other, that the common air may have no access. You then alternately inspire, and expire the gas, till you perceive its effects. But I cannot consent to your making the experiment; for the nerves are sometimes unpleasantly affected by it, and I would not run any risk of that kind.

EMILY.

I should like, at least, to see somebody breathe it; but pray by what means is this curious gas obtained?

MRS. B.

It is procured from *nitrat of ammonia*, an artificial salt which yields this gas on the application of a gentle heat. I have put some of the salt into a retort, and by the aid of a lamp the gas will be extricated. —

CAROLINE.

Bubbles of air begin to escape through the neck of the retort into the water apparatus; will you not collect them?

MRS. B.

The gas that first comes over need not be preserved, as it consists of little more than the common air that was in the retort; besides, there is always in this experiment a quantity of watery vapour which must come away before the nitrous oxyd appears.

EMILY.

Watery vapour! Whence does that proceed? There is no water in nitrat of ammonia?

MRS. B.

You must recollect that there is in every salt a quantity of water of crystallisation, which may be evaporated by heat alone. But, besides this, water is actually generated in this experiment, as you will see presently. First tell me, what are the constituent parts of nitrat of ammonia?

EMILY.

Ammonia, and nitric acid: this salt, therefore, contains three different elements, nitrogen and hydrogen, which produce the ammonia; and oxygen, which, with nitrogen, forms the acid.

MRS. B.

Well then, in this process the ammonia is decomposed; the hydrogen quits the nitrogen to combine with some of the oxygen of the nitric acid, and forms with it the watery vapour which is now coming over. When that is effected, what will you expect to find?

EMILY.

Nitrous acid instead of nitric acid, and nitrogen instead of ammonia.

MRS. B.

Exactly so; and the nitrous acid and nitrogen combine, and form the gaseous oxyd of nitrogen, in which the proportion of oxygen is 37 parts to 63 of nitrogen.

You may have observed, that for a little while

no bubbles of air have come over, and we have perceived only a stream of vapour condensing as it issued into the water. — Now bubbles of air again make their appearance, and I imagine that by this time all the watery vapour is come away, and that we may begin to collect the gas. We may try whether it is pure, by filling a phial with it, and plunging a taper into it — yes, it will do now, for the taper burns brighter than in the common air, and with a greenish flame.

CAROLINE.

But how is that? I thought no gas would support combustion but oxygen or chlorine.

MRS. B.

Or any gas that contains oxygen, and is ready to yield it, which is the case with this in a considerable degree; it is not, therefore, surprising that it should accelerate the combustion of the taper.

You see that the gas is now produced in great abundance; we shall collect a large quantity of it, and I dare say that we shall find some of the family who will be curious to make the experiment of respiring it. Whilst this process is going on, we may take a general survey of the most important combinations of the nitric and nitrous acids with the alkalies.

The first of these is *nitrat of potash*, commonly called *nitre* or *saltpetre.*

CAROLINE.

Is not that the salt with which gunpowder is made?

MRS. B.

Yes. Gunpowder is a mixture of five parts of nitre to one of sulphur, and one of charcoal. — Nitre from its great proportion of oxygen, and from the facility with which it yields it, is the basis of most detonating compositions.

EMILY.

But what is the cause of the violent detonation of gunpowde when set fire to?

MRS. B.

Detonation may proceed from two causes; the sudden formation or destruction of an elastic fluid. In the first case, when either a solid or liquid is instantaneously converted into an elastic fluid, the prodigious and sudden expansion of the body strikes the air with great violence, and this concussion produces the sound called detonation.

CAROLINE.

That I comprehend very well; but how can a similar effect be produced by the destruction of a gas?

MRS. B.

A gas can be destroyed only by condensing it to a liquid or solid state; when this takes place suddenly, the gas, in assuming a new and more compact form, produces a vacuum, into which the surrounding air rushes with great impetuosity; and it is by that rapid and violent motion that the sound is produced. In all detonations, therefore, gases are either suddenly formed, or destroyed. In that of gunpowder, can you tell me which of these two circumstances takes place?

EMILY.

As gunpowder is a solid, it must, of course, produce the gases in its detonation; but how, I cannot tell.

MRS. B.

The constituents of gunpowder, when heated to a certain degree, enter into a number of new combinations, and are instantaneously converted into a variety of gases, the sudden expansion of which gives rise to the detonation.

CAROLINE.

And in what instance does the destruction or condensation of gases produce detonation?

MRS. B.

I can give you one with which you are well

acquainted; the sudden combination of the oxygen and hydrogen gases.

CAROLINE.

True; I recollect perfectly that hydrogen detonates with oxygen when the two gases are converted into water.

MRS. B.

But let us return to the nitrat of potash.—This salt is decomposed when exposed to heat, and mixed with any combustible body, such as carbon, sulphur, or metals, these substances oxydating rapidly at the expense of the nitrat. I must show you an instance of this.—I expose to the fire some of the salt in a small iron ladle, and, when it is sufficiently heated, add to it some powdered charcoal; this will attract the oxygen from the salt, and be converted into carbonic acid.—

EMILY.

But what occasions that crackling noise, and those vivid flashes that accompany it?

MRS. B.

The rapidity with which the carbonic acid gas is formed occasions a succession of small detonations, which, together with the emission of flame, is called *deflagration*.

Nitrat of ammonia we have already noticed,

on account of the gaseous oxyd of nitrogen which is obtained from it.

Nitrat of silver is the lunar caustic, so remarkable for its property of destroying animal fibre, for which purpose it is often used by surgeons. — We have said so much on a former occasion, on the mode in which caustics act on animal matter, that I shall not detain you any longer on this subject.

We now come to the CARBONIC ACID, which we have already had many opportunities of noticing. You recollect that this acid may be formed by the combustion of carbon, whether in its imperfect state of charcoal, or in its purest form of diamond. And it is not necessary, for this purpose, to burn the carbon in oxygen gas, as we did in the preceding lecture; for you need only light a piece of charcoal and suspend it under a receiver on the water bath. The charcoal will soon be extinguished, and the air in the receiver will be found mixed with carbonic acid. The process, however, is much more expeditious if the combustion be performed in pure oxygen gas.

CAROLINE.

But how can you separate the carbonic acid,

obtained in this manner, from the air with which it is mixed?

MRS. B.

The readiest mode is to introduce under the receiver a quantity of caustic lime, or caustic alkali, which soon attracts the whole of the carbonic acid to form a carbonat.—The alkali is found increased in weight, and the volume of the air is diminished by a quantity equal to that of the carbonic acid which was mixed with it.

EMILY.

Pray is there no method of obtaining pure carbon from carbonic acid?

MRS. B.

For a long time it was supposed that carbonic acid was not decompoundable; but Mr. Tennant discovered, a few years ago, that this acid may be decomposed by burning phosphorus in a closed vessel with carbonat of soda or carbonat of lime: the phosphorus absorbs the oxygen from the carbonat, whilst the carbon is separated in the form of a black powder. This decomposition, however, is not effected simply by the attraction of the phosphorus for oxygen, since it is weaker than that of charcoal; but the attraction of the alkali or lime for the phosphoric acid, unites its power at the same time.

CAROLINE.

Cannot we make that experiment?

MRS. B.

Not easily; it requires being performed with extreme nicety, in order to obtain any sensible quantity of carbon, and the experiment is much too delicate for me to attempt it. But there can be no doubt of the accuracy of Mr. Tennant's results; and all chemists now agree, that one hundred parts of carbonic acid gas consists of about twenty-eight parts of carbon to seventy-two of oxygen gas. But if you recollect, we decomposed carbonic acid gas the other day by burning potassium in it.

CAROLINE.

True, so we did; and found the carbon precipitated on the regenerated potash.

MRS. B.

Carbonic acid gas is found very abundantly in nature; it is supposed to form about one thousandth part of the atmosphere, and is constantly produced by the respiration of animals; it exists in a great variety of combinations, and is exhaled from many natural decompositions. It is contained in a state of great purity in certain caves, such as the *Grotto del Cane*, near Naples.

EMILY.

I recollect having read an account of that grotto, and of the cruel experiments made on the poor dogs, to gratify the curiosity of strangers. But I understood that the vapour exhaled by this cave was called *fixed air*.

MRS. B.

That is the name by which carbonic acid was known before its chemical composition was discovered.—This gas is more destructive of life than any other; and if the poor animals that are submitted to its effects are not plunged into cold water as soon as they become senseless, they do not recover. It extinguishes flame instantaneously. I have collected some in this glass, which I will pour over the candle.

CAROLINE.

This is extremely singular — it seems to extinguish it as it were by enchantment, as the gas is invisible. I never should have imagined that gas could have been poured like a liquid.

MRS. B.

It can be done with carbonic acid only, as no other gas is sufficiently heavy to be susceptible of being poured out in the atmospherical air without mixing with it.

EMILY.

Pray by what means did you obtain this gas?

MRS. B.

I procured it from marble. Carbonic acid gas has so strong an attraction for all the alkalies and alkaline earths, that these are always found in nature in the state of carbenats. Combined with lime, this acid forms chalk, which may be considered as the basis of all kinds of marbles, and calcareous stones. From these substances carbonic acid is easily separated, as it adheres so slightly to its combinations, that the carbonats are all decomposable by any of the other acids. I can easily show you how I obtained this gas; I poured some diluted sulphuric acid over pulverised marble in this bottle (the same which we used the other day to prepare hydrogen gas), and the gas escaped through the tube connected with it; the operation still continues, as you may easily perceive —

EMILY.

Yes, it does; there is a great fermentation in the glass vessel. What singular commotion is excited by the sulphuric acid taking possession of the lime, and driving out the carbonic acid!

CAROLINE.

But did the carbonic acid exist in a gaseous state in the marble?

MRS. B.

Certainly not; the acid, when in a state of combination, is capable of existing in a solid form.

CAROLINE.

Whence, then, does it obtain the caloric necessary to convert it into gas?

MRS. B.

It may be supplied in this case from the mixture of sulphuric acid and water, which produces an evolution of heat, even greater than is required for the purpose; since, as you may perceive by touching the glass vessel, a considerable quantity of the caloric disengaged becomes sensible. But a supply of caloric may be obtained also from a diminution of capacity for heat, occasioned by the new combination which takes place; and, indeed, this must be the case when other acids are employed for the disengagement of carbonic acid gas, which do not, like the sulphuric, produce heat on being mixed with water. Carbonic acid may likewise be disengaged from its combinations by heat alone, which restores it to its gaseous state.

CAROLINE.

It appears to me very extraordinary that the same gas, which is produced by the burning of wood and coals, should exist also in such bodies as marble, and chalk, which are incombustible substances.

MRS. B.

I will not answer that objection, Caroline, because I think I can put you in a way of doing it yourself. Is carbonic acid combustible?

CAROLINE.

Why, no — because it is a body that has been already burnt; it is carbon only, and not the acid, that is combustible.

MRS. B.

Well, and what inference do you draw from this?

CAROLINE.

That carbonic acid cannot render the bodies with which it is united combustible; but that simple carbon does, and that it is in this elementary state that it exists in wood, coals, and a great variety of other combustible bodies. — Indeed, Mrs. B., you are very ungenerous; you are not satisfied with convincing me that my objections are frivolous, but you oblige me to prove them so myself.

MRS. B.

You must confess, however, that I make ample amends for the detection of error, when I enable you to discover the truth. You, understand, now, I hope, that carbonic acid is equally produced by the decomposition of chalk, or by the combustion of charcoal. These processes are certainly of a very different nature; in the first case the acid is already formed, and requires nothing more than heat to restore it to its gaseous state; whilst, in the latter, the acid is actually made by the process of combustion.

CAROLINE.

I understand it now perfectly. But I have just been thinking of another difficulty, which, I hope, you will excuse my not being able to remove myself. How does the immense quantity of calcareous earth, which is spread all over the globe, obtain the carbonic acid with which it is combined?

MRS. B.

The question is, indeed, not very easy to answer; but I conceive that the general carbonisation of calcareous matter may have been the effect of a general combustion, occasioned by some revolution of our globe, and producing an immense supply of carbonic acid, with which the calcareous matter became impregnated; or that this may have been effected by a gradual absorption of carbonic acid from the atmosphere. — But this would lead us to discussions which we cannot indulge in, without deviating too much from our subject.

EMILY.

How does it happen that we do not perceive the pernicious effects of the carbonic acid which is floating in the atmosphere?

MRS. B.

Because of the state of very great dilution in which it exists there. But can you tell me, Emily, what are the sources which keep the atmosphere constantly supplied with this acid?

EMILY.

I suppose the combustion of wood, coals, and other substances, that contain carbon.

MRS. B.

And also the breath of animals.

CAROLINE.

The breath of animals! I thought you said that this gas was not at all respirable, but on the contrary, extremely poisonous.

MRS. B.

So it is; but although animals cannot breathe in carbonic acid gas, yet, in the process of respiration, they have the power of forming this gas in their lungs; so that the air which we *expire*, or reject from the lungs, always contains a certain proportion of carbonic acid, which is much greater than that which is commonly found in the atmosphere.

CAROLINE.

But what is it that renders carbonic acid such a deadly poison?

MRS. B.

The manner in which this gas destroys life, seems to be merely by preventing the access of respirable air; for carbonic acid gas, unless very much diluted with common air, does not penetrate into the lungs, as the windpipe actually contracts and refuses it admittance. — But we must dismiss this subject at present, as we shall have an opportunity of treating of respiration much more fully, when we come to the chemical functions of animals.

EMILY.

Is carbonic acid as destructive to the life of vegetables as it is to that of animals?

MRS. B.

If a vegetable be completely immersed in it, I believe it generally proves fatal to it; but mixed in certain proportions with atmospherical air, it is, on the contrary, very favourable to vegetation.

You remember, I suppose, our mentioning the mineral waters, both natural and artificial, which contain carbonic acid gas?

CAROLINE.

You mean the Seltzer water?

MRS. B.

That is one of those which are the most used; there are, however, a variety of others into which carbonic acid enters as an ingredient: all these waters are usually distinguished by the name of *acidulous* or *gaseous mineral waters.*

The class of salts called *carbonats* is the most numerous in nature; we must pass over them in a very cursory manner, as the subject is far too extensive for us to enter on it in detail. The state of carbonat is the natural state of a vast number of minerals, and particularly of the alkalies and alkaline earths, as they have so great an attraction for the carbonic acid, that they are almost always found combined with it; and you may recollect that it is only by separating them from this acid, that they acquire that causticity and those striking qualities which I have formerly described. All marbles, chalks, shells, calcareous spars, and limestones of every description, are neutral salts, in which *lime*, their common basis, has lost all its characteristic properties.

EMILY.

But if all these various substances are formed by the union of lime with carbonic acid, whence arises their diversity of form and appearance?

MRS. B.

Both from the different proportions of their component parts, and from a variety of foreign ingredients which may be occasionally blended with them: the veins and colours of marbles, for instance, proceed from a mixture of metallic substances; silex and alumine also frequently enter into these combinations. The various carbonats, therefore, that I have enumerated, cannot be considered as pure unadulterated neutral salts, although they certainly belong to that class of bodies.

CONVERSATION XIX.

ON THE BORACIC, FLUORIC, MURIATIC, AND OXYGENATED MURIATIC ACIDS; AND ON MURIATS. — ON IODINE AND IODIC ACID.

MRS. B.

We now come to the three remaining acids with simple bases, the compound nature of which, though long suspected, has been but recently proved. The chief of these is the muriatic; but I shall first describe the two others, as their bases have been obtained more distinctly than that of the muriatic acid.

You may recollect I mentioned the BORACIC ACID. This is found very sparingly in some parts of Europe, but for the use of manufactures we have always received it from the remote country of Thibet, where it is found in some lakes, combined with soda. It is easily separated from the soda by sulphuric acid, and appears in the form of shining scales, as you see here.

CAROLINE.

I am glad to meet with an acid which we need

not be afraid to touch; for I perceive, from your keeping it in a piece of paper, that it is more innocent than our late acquaintance, the sulphuric and nitric acids.

MRS. B.

Certainly; but being more inert, you will not find its properties so interesting. However, its decomposition, and the brilliant spectacle it affords when its basis again unites with oxygen, atones for its want of other striking qualities.

Sir H. Davy succeeded in decomposing the boracic acid, (which had till then been considered as undecompoundable,) by various methods. On exposing this acid to the Voltaic battery, the positive wire gave out oxygen, and on the negative wire was deposited a black substance, in appearance resembling charcoal. This was the basis of the acid, which Sir H. Davy has called *Boracium*, or *Boron*.

The same substance was obtained in more considerable quantities, by exposing the acid to a great heat in an iron gun-barrel.

A third method of decomposing the boracic acid consisted in burning potassium in contact with it in vacuo. The potassium attracts the oxygen from the acid, and leaves its basis in a separate state.

The recomposition of this acid I shall show

you, by burning some of its basis, which you see here, in a retort full of oxygen gas. The heat of a candle is all that is required for this combustion. —

EMILY.

The light is astonishingly brilliant, and what beautiful sparks it throws out!

MRS. B.

The result of this combustion is the boracic acid, the nature of which, you see, is proved both by analytic and synthetic means. Its basis has not, it is true, a metallic appearance; but it makes very hard alloys with other metals.

EMILY.

But pray, Mrs. B., for what purpose is the boracic acid used in manufactures?

MRS. B.

Its principal use is in conjunction with soda, that is, in the state of *borat of soda*, which in the arts is commonly called borax. This salt has a peculiar power of dissolving metallic oxyds, and of promoting the fusion of substances capable of being melted; it is accordingly employed in various metallic arts; it is used, for example, to remove the oxyd from the surface of metals, and is often employed in the assaying of metallic ores.

Let us now proceed to the FLUORIC ACID. This acid is obtained from a substance which is found frequently in mines, and particularly in those of Derbyshire, called *fluor*, a name which it acquired from the circumstance of its being used to render the ores of metals more fluid when heated.

CAROLINE.

Pray is not this the Derbyshire spar, of which so many ornaments are made?

MRS. B.

The same; but though it has long been employed for a variety of purposes, its nature was unknown until Scheele, the great Swedish chemist, discovered that it consisted of lime united with a peculiar acid, which obtained the name of *fluoric acid.* It is easily separated from the lime by the sulphuric acid, and unless condensed in water, ascends in the form of gas. A very peculiar property of this acid is its union with siliceous earths, which I have already mentioned. If the distillation of this acid is performed in glass vessels, they are corroded, and the siliceous part of the glass comes over, united with the gas; if water is then admitted, part of the silex is deposited, as you may observe in this jar.

CAROLINE.

I see white flakes forming on the surface of the water; is that silex?

MRS. B.

Yes it is. This power of corroding glass has been used for engraving, or rather etching, upon it. The glass is first covered with a coat of wax, through which the figures to be engraved are to be scratched with a pin; then pouring the fluoric acid over the wax, it corrodes the glass where the scratches have been made.

CAROLINE.

I should like to have a bottle of this acid, to make engravings.

MRS. B.

But you could not have it in a *glass* bottle, for in that case the acid would be saturated with silex, and incapable of executing an engraving; the same thing would happen were the acid kept in vessels of porcelain or earthen-ware; this acid must therefore be both prepared and preserved in vessels of silver.

If it be distilled from fluor spar and vitriolic acid, in silver or leaden vessels, the receiver being kept very cold during the distillation, it assumes the form of a dense fluid, and in that state is the most intensely corrosive substance known. This seems to be the acid combined with a little water. It may be called *hydro-fluoric acid;* and Sir H. Davy has been led, from some late experiments on the subject, to consider *pure* fluoric acid as a compound of a certain unknown principle, which he calls *fluorine*, with hydrogen.

Sir H. Davy has also attempted to decompose the fluoric acid by burning potassium in contact with it; but he has not yet been able by this or any other method, to obtain its basis in a distinct separate state.

We shall conclude our account of the acids with that of the MURIATIC ACID, which is perhaps the most curious and interesting of all of them. It is found in nature combined with soda, lime, and magnesia. *Muriat of soda* is the com mon sea-salt, and from this substance the acid is usually disengaged by means of the sulphuric acid. The natural state of the muriatic acid is that of an invisible permanent gas, at the common temperature of the atmosphere; but it has a remarkably strong attraction for water, and assumes the form of a whitish cloud whenever it meets any moisture to combine with. This acid is remarkable for its peculiar and very pungent smell, and possesses, in a pòwerful degree, most of the acid properties. Here is a bottle containing muriatic acid in a liquid state.

CAROLINE.

And how is it liquefied?

MRS. B.

By impregnating water with it; its strong attraction for water makes it very easy to obtain it in a liquid form. Now, if I open the phial, you may observe a kind of vapour rising from it, which is muriatic acid gas, of itself invisible, but made apparent by combining with the moisture of the atmosphere.

EMILY.

Have you not any of the pure muriatic acid gas?

MRS. B.

This jar is full of that acid in its gaseous state — it is inverted over mercury instead of water, because, being absorbable by water, this gas cannot be confined by it. — I shall now raise the jar a little on one side, and suffer some of the gas to escape. — You see that it immediately becomes visible in the form of a cloud.

EMILY.

It must be, no doubt, from its uniting with the moisture of the atmosphere, that it is converted into this dewy vapour.

MRS. B.

Certainly; and for the same reason, that is to say, its extreme eagerness to unite with water, this gas will cause snow to melt as rapidly as an intense fire.

This acid proved much more refractory when Sir H. Davy attempted to decompose it than the other two undecompounded acids. It is singular that potassium will burn in muriatic acid, and be converted into potash, without decomposing the acid, and the result of this combustion is a *muriat of potash*; for the potash, as soon as it is regenerated, combines with the muriatic acid.

CAROLINE.

But how can the potash be regenerated if the muriatic acid does not oxydate the potassium?

MRS. B.

The potassium, in this process, obtains oxygen from the moisture with which the muriatic acid is always combined, and accordingly hydrogen, resulting from the decomposition of the moisture, is invariably evolved.

EMILY.

But why not make these experiments with dry muriatic acid?

MRS. B.

Dry acids cannot be acted on by the Voltaic battery, because acids are non-conductors of electricity, unless moistened. In the course of a number of experiments which Sir H. Davy made upon acids in a state of dryness, he observed that the presence of water appeared always necessary to develop the acid properties, so that acids are not even capable of reddening vegetable blues if they have been carefully deprived of moisture. This remarkable circumstance led him to suspect, that water, instead of oxygen, may be the acidifying principle; but this he threw out rather as a conjecture than as an established point.

Sir H. Davy obtained very curious results from burning potassium in a mixture of phosphorus and muriatic acid, and also of sulphur and muriatic acid; the latter detonates with great violence. All his experiments, however, failed in presenting to his view the basis of the muriatic acid, of which he was in search; and he was at last induced to form an opinion respecting the nature of this acid, which I shall presently explain.

EMILY.

Is this acid susceptible of different degrees of oxygenation?

MRS. B.

Yes, for though we cannot deoxygenate this acid, yet we may add oxygen to it.

CAROLINE.

Why, then, is not the least degree of oxygenation of the acid called the *muriatous*, and the higher degree the *muriatic* acid?

MRS. B.

Because, instead of becoming, like other acids, more dense, and more acid by an addition of oxygen, it is rendered on the contrary more volatile, more pungent, but less acid, and less absorbable by water. These circumstances, therefore, seem to indicate the propriety of making an exception to the nomenclature. The highest degree of oxygenation of this acid has been distinguished by the additional epithet of *oxygenated*, or, for the sake of brevity, *oxy*, so that it is calléd the *oxygenated*, or *oxy-muriatic acid*. This likewise exists in a gaseous form, at the temperature of the atmosphere; it is also susceptible of being absorbed by water, and can be congealed, or solidified, by a certain degree of cold.

EMILY.

And how do you obtain the oxy-muriatic acid?

MRS. B.

In various ways; but it may be most conveniently obtained by distilling liquid muriatic acid over oxyd of manganese, which supplies the acid with the additional oxygen. One part of the acid being put into a retort, with two parts of the oxyd of manganese, and the heat of a lamp applied, the gas is soon disengaged, and may be received over water, as it is but sparingly absorbed by it. — I have collected some in this jar —

CAROLINE.

It is not invisible, like the generality of gases; for it is of a yellowish colour.

MRS. B.

The muriatic acid extinguishes flame, whilst, on the contrary, the oxy-muriatic makes the flame larger, and gives it a dark red colour. Can you account for this difference in the two acids?

EMILY.

Yes, I think so; the muriatic acid will not supply the flame with the oxygen necessary for its support; but when this acid is further oxygenated, it will part with its additional quantity of oxygen, and in this way support combustion.

MRS. B.

That is exactly the case; indeed the oxygen added to the muriatic acid, adheres so slightly to it, that it is separated by mere exposure to the sun's rays. This acid is decomposed also by combustible bodies, many of which it burns, and actually inflames, without any previous increase of temperature.

CAROLINE.

That is extraordinary, indeed! I hope you mean to indulge us with some of these experiments?

MRS. B.

I have prepared several glass jars of oxy-muriatic acid gas for that purpose. In the first we shall introduce some Dutch gold leaf. — Do you observe that it takes fire?

EMILY.

Yes, indeed it does — how wonderful it is! It became immediately red hot, but was soon smothered in a thick vapour.

CAROLINE.

What a disagreeable smell!

MRS. B.

We shall try the same experiment with phosphorus in another jar of this acid.—You had better keep your handkerchief to your nose when I open it—now let us drop into it this little piece of phosphorus —

CAROLINE.

It burns really; and almost as brilliantly as in oxygen gas! But, what is most extraordinary, these combustions take place without the metal or phosphorus being previously lighted, or even in the least heated.

MRS. B.

All these curious effects are owing to the very great facility with which this acid yields oxygen to such bodies as are strongly disposed to combine with it. It appears extraordinary indeed to see bodies, and metals in particular, melted down and inflamed, by a gas without any increase of temperature, either of the gas, or of the combustible. The phenomenon, however, is, you see, well accounted for.

EMILY.

Why did you burn a piece of Dutch gold leaf rather than a piece of any other metal?

MRS. B.

Because, in the first place, it is a composition

of metals (consisting chiefly of copper) which burns readily; and I use a thin metallic leaf in preference to a lump of metal, because it offers to the action of the gas but a small quantity of matter under a large surface. Filings, or shavings, would answer the purpose nearly as well; but a lump of metal, though the surface would oxydate with great rapidity, would not take fire. Pure gold is not inflamed by oxy-muriatic acid gas, but it is rapidly oxydated, and dissolved by it; indeed, this acid is the only one that will dissolve gold.

EMILY.

This, I suppose, is what is commonly called *aqua regia*, which you know is the only thing that will act upon gold.

MRS. B.

That is not exactly the case either; for aqua regia is composed of a mixture of muriatic acid and nitric acid. — But, in fact, the result of this mixture is the formation of oxy-muriatic acid, as the muriatic acid oxygenates itself at the expence of the nitric; this mixture, therefore, though it bears the name of *nitro-muriatic acid*, acts on gold merely in virtue of the oxy-muriatic acid which it contains.

Sulphur, volatile oils, and many other substances, will burn in the same manner in oxy-muriatic acid gas; but I have not prepared a sufficient quantity of it, to show you the combustion of all these bodies.

CAROLINE.

There are several jars of the gas yet remaining.

MRS. B.

We must reserve these for future experiments. The oxy-muriatic acid does not, like other acids, redden the blue vegetable colours; but it totally destroys any colour, and turns all vegetables perfectly white. Let us collect some vegetable substances to put into this glass, which is full of gas.

EMILY.

Here is a sprig of myrtle —

CAROLINE.

And here some coloured paper —

MRS. B.

We shall also put in this piece of scarlet riband, and a rose —

EMILY.

Their colours begin to fade immediately! But how does the gas produce this effect?

MRS. B.

The oxygen combines with the colouring matter of these substances, and destroys it; that is to say, destroys the property which these colours had of reflecting only one kind of rays, and renders them capable of reflecting them all, which, you know, will make them appear white. Old prints may be cleaned by this acid, for the paper will be whitened without injury to the impression, as printer's ink is made of materials (oil and lamp black) which are not acted upon by acids.

This property of the oxy-muriatic acid has lately been employed in manufactures in a variety of bleaching processes; but for these purposes the gas must be dissolved in water, as the acid is thus rendered much milder and less powerful in its effects; for, in a gaseous state, it would destroy the texture, as well as the colour of the substance submitted to its action.

CAROLINE.

Look at the things which we put into the gas they have now entirely lost their colour!

MRS. B.

The effect of the acid is almost completed; and, if we were to examine the quantity that remains, we should find it to consist chiefly of muriatic acid.

The oxy-muriatic acid has been used to purify the air in fever hospitals and prisons, as it burns and destroys putrid effluvia of every kind. The infection of the small-pox is likewise destroyed by this gas, and matter that has been submitted to its influence will no longer generate that disorder.

CAROLINE.

Indeed, I think the remedy must be nearly as bad as the disease; the oxy-muriatic acid has such a dreadfully suffocating smell.

MRS. B.

It is certainly extremely offensive; but by keeping the mouth shut, and wetting the nostrils with liquid ammonia, in order to neutralize the vapour as it reaches the nose, its prejudicial effects may be in some degree prevented. At any rate, however, this mode of disinfection can hardly be used in places that are inhabited. And as the vapour of nitric acid, which is scarcely less efficacious for this purpose, is not at all prejudicial, it is usually preferred on such occasions.

CAROLINE.

You have not told us yet what is Sir H. Davy's new opinion respecting the nature of muriatic acid, to which you alluded a few minutes ago?

MRS. B.

True; I avoided noticing it then, because you could not have understood it without some previous knowledge of the oxy-muriatic acid, which I have but just introduced to your acquaintance.

Sir H. Davy's idea is that muriatic acid, instead of being a compound, consisting of an unknown basis and oxygen, is formed by the union of oxy-muriatic gas with hydrogen.

EMILY.

Have you not told us just now that oxy-muriatic gas was itself a compound of muriatic acid and oxygen?

MRS. B.

Yes; but according to Sir H. Davy's hypothesis, oxy-muriatic gas is considered as a simple body, which contains no oxygen — as a substance of its own kind, which has a great analogy to oxygen in most of its properties, though in others it differs entirely from it. — According to this view of the subject, the name of *oxy-muriatic acid* can no longer be proper, and therefore Sir H. Davy has adopted that of *chlorine*, or *chlorine gas*, a name which is simply expressive of its greenish colour; and in compliance with that philosopher's theory, we have placed chlorine in our table among the simple bodies.

CAROLINE.

But what was Sir H. Davy's reason for adopting an opinion so contrary to that which had hitherto prevailed?

MRS. B.

There are many circumstances which are favourable to the new doctrine; but the clearest and simplest fact in its support is, that if hydrogen gas and oxy-muriatic gas be mixed together, both these gases disappear, and muriatic acid gas is formed.

EMILY.

That seems to be a complete proof; is it not considered as perfectly conclusive?

MRS. B.

Not so decisive as it appears at first sight; because it is argued by those who still incline to the old doctrine, that muriatic cid gas, however dry it may be, always contains a certain quantity of water, which is supposed essential to its formation. So that, in the experiment just mentioned, this water is supplied by the union of the hydrogen gas with the oxygen of the oxy-muriatic acid; and therefore the mixture resolves itself into the base of muriatic acid and water, that is, muriatic acid gas.

CAROLINE.

I think the old theory must be the true one; for otherwise how could you explain the formation of oxy-muriatic gas, from a mixture of muriatic acid and oxyd of manganese?

MRS. B.

Very easily; you need only suppose that in this process the muriatic acid is decomposed; its hydrogen unites with the oxygen of the manganese to form water, and the chlorine appears in its separate state.

EMILY.

But how can you explain the various combustions which take place in oxy-muriatic gas, if you consider it as containing no oxygen?

MRS. B.

We need nly suppose that combustion is the result of intense chemical action; so that chlorine, like oxygen, in combining with bodies, forms compounds which have less capacity for caloric than their constituent principles, and, therefore, caloric is evolved at the moment of their combination.

EMILY.

If, then, we may explain every thing by either theory, to which of the two shall we give the preference?

MRS. B.

It will, perhaps, be better to wait for more positive proofs, if such can be obtained, before we decide positively upon the subject. The new doctrine has certainly gained ground very rapidly, and may be considered as nearly established; but seveal competent judges still refuse their assent to it, and until that theory is very generally adopted, it may be as well for us still occasionally to use the language to which chemists have long been accustomed. But let us proceed to the examination of salts formed by muriatic acid.

Among the compound salts formed by muriatic acid, the *muriat of soda*, or common salt, is the most interesting * The uses and properties of this salt are too well known to require much comment. Besides the pleasant flavour it imparts to the food, it is very wholesome, when not used to excess, as it assists the process of digestion.

Sea-water is the great source from which muriat of soda is extracted by evaporation. But it is also found in large solid masses in the bowels of the earth, in England, and in many other parts of the world.

EMILY.

I thought that salts, when solid, were always in the state of crystals; but the common table-salt is in the form of a coarse white powder.

* According to Sir H. Davy's views of the nature of the muriatic and oxy-muriatic acids, dry muriat of soda is a compound of sodium and chlorine, for it may be formed by the direct combination of oxy-muriatic gas and sodium. In his opinion, therefore, what we commonly call muriat of soda contains neither soda nor muriatic acid.

MRS. B.

Crystallisation depends, as you may recollect, on the slow and regular reunion of particles dissolved in a fluid; common sea-salt is only in a state of imperfect crystallisation, because the process by which it is prepared is not favourable to the formation of regular crystals. But if you dissolve it, and afterwards evaporate the water slowly, you will obtain a regular crystallisation.

Muriat of ammonia is another combination of this acid, which we have already mentioned as the principal source from which ammonia is derived.

I can at once show you the formation of this salt by the immediate combination of muriatic acid with ammonia.—These two glass jars contain, the one muriatic acid gas, the other ammoniacal gas, both of which are perfectly invisible—now, if I mix them together, you see they immediately form an opake white cloud, like smoke.—If a thermometer was placed in the jar in which these gases are mixed, you would perceive that some heat is at the same time produced.

EMILY.

The effects of chemical combinations are, indeed, wonderful!—How extraordinary it is that two invisible bodies should become visible by their union!

MRS. B.

This strikes you with astonishment, because it is a phenomenon which nature seldom exhibits to our view; but the most common of her operations are as wonderful, and it is their frequency only that prevents our regarding them with equal admiration. What would be more surprising, for instance, than combustion, were it not rendered so familiar by custom?

EMILY.

That is true.—But pray, Mrs. B., is this white cloud the salt that produces ammonia? How different it is from the solid muriat of ammonia which you once showed us!

MRS. B.

It is the same substance which first appears in the state of vapour, but will soon be condensed by cooling against the sides of the jar, in the form of very minute crystals.

We may now proceed to the *oxy-muriats*. In this class of salts the *oxy-muriat of potash* is the most worthy of our attention, for its striking properties. The acid, in this state of combination, contains a still greater proportion of oxygen than when alone.

CAROLINE.

But how can the oxy-muriatic acid acquire an increase of oxygen by combining with potash?

MRS. B.

It does not really acquire an additional quantity of oxygen, but it loses some of the muriatic acid, which produces the same effect, as the acid which remains is proportionably super-oxygenated.*

If this salt be mixed, and merely rubbed together with sulphur, phosphorus, charcoal, or indeed any other combustible, it explodes strongly.

CAROLINE.

Like gun-powder, I suppose, it is suddenly converted into elastic fluids?

MRS. B.

Yes; but with this remarkable difference, that no increase of temperature, any further than is produced by gentle friction, is required in this instance. Can you tell me what gases are generated by the detonation of this salt with charcoal?

EMILY.

Let me consider . . . . . The oxy-muriatic acid parts with its excess of oxygen to the charcoal, by which means it is converted into muriatic acid gas; whilst the charcoal, being burnt by the oxygen, is

* According to Sir H. Davy's new views, just explained, oxy-muriat of potash is a compound of chlorine with oxyd of potassium.

changed to carbonic acid gas — What becomes of the potash I cannot tell.

MRS. B.

That is a fixed product which remains in the vessel.

CAROLINE.

But since the potash does not enter into the new combinations, I do not understand of what use it is in this operation. Would not the oxy-muriatic acid and the charcoal produce the same effect without it?

MRS. B.

No; because there would not be that very great concentration of oxygen which the combination with the potash produces, as I have just explained.

I mean to show you this experiment, but I would advise you not to repeat it alone; for if care be not taken to mix only very small quantities at a time, the detonation will be extremely violent, and may be attended with dangerous effects. You see I mix an exceedingly small quantity of the salt with a little powdered charcoal, in this Wedgwood mortar, and rub them together with the pestle —

CAROLINE.

Heavens! How can such a loud explosion be produced by so small a quantity of matter?

MRS. B.

You must consider that an extremely small quantity of solid substance may produce a very great volume of gases; and it is the sudden evolution of these which occasions the sound.

EMILY.

Would not oxy-muriat of potash make stronger gunpowder than nitrat of potash?

MRS. B.

Yes; but the preparation, as well as the use of this salt, is attended with so much danger, that it is never employed for that purpose.

CAROLINE.

There is no cause to regret it, I think; for the common gunpowder is quite sufficiently destructive.

MRS. B.

I can show you a very curious experiment with this salt; but it must again be on condition that you will never attempt to repeat it by yourselves. I throw a small piece of phosphorus into this glass of water; then a little oxy-muriat of potash; and, lastly, I pour in (by means of this funnel, so as to bring it in contact with the two other ingredients at the bottom of the glass) a small quantity of sulphuric acid —

CAROLINE.

This is, indeed, a beautiful experiment! The phosphorus takes fire and burns from the bottom of the water.

EMILY.

How wonderful it is to see flame bursting out under water, and rising through it! Pray, how is this accounted for?

MRS. B.

Cannot you find it out, Caroline?

EMILY.

Stop — I think I can explain it. Is it not because the sulphuric acid decomposes the salt by combining with the potash, so as to liberate the oxy-muriatic acid gas by which the phosphoric is set on fire?

MRS. B.

Very well, Emily; and with a little more reflection you would have discovered another concurring circumstance, which is, that an increase of temperature is produced by the mixture of the sulphuric acid and water, which assists in promoting the combustion of the phosphorus.

I must, before we part, introduce to your acquaintance the newly-discovered substance IODINE, which you may recollect we placed next to oxygen and chlorine in our table of simple bodies.

CAROLINE.

Is this also a body capable of maintaining combustion like oxygen and chlorine?

MRS. B.

It is; and although it does not so generally disengage light and heat from inflammable bodies, as oxygen and chlorine do, yet it is capable of combining with most of them; and sometimes, as in the instance of potassium and phosphorus, the combination is attended with an actual appearance of light and heat.

CAROLINE.

But what sort of a substance is iodine: what is its form, and colour?

MRS. B.

It is a very singular body, in many respects. At the ordinary temperature of the atmosphere, it commonly appears in the form of blueish black crystalline scales, such as you see in this tube.

CAROLINE.

They shine like black lead, and some of the scales have the shape of lozenges.

MRS. B.

That is actually the form which the crystals of iodine often assume. But if we heat them gently, by holding the tube over the flame of a candle, see what a change takes place in them.

CAROLINE.

How curious! They seem to melt, and the tube immediately fills with a beautiful violet vapour. But look, Mrs. B., the same scales are now appearing at the other end of the tube.

MRS. B.

This is in fact a sublimation of iodine, from one part of the tube to another; but with this remarkable peculiarity, that, while in the gaseous state, iodine assumes that bright violet colour, which, as you may already perceive, it loses as the tube cools, and the substance resumes its usual solid form. — It is from the violet colour of the gas that iodine has obtained its name.

CAROLINE.

But how is this curious substance obtained?

MRS. B.

It is found in the ley of ashes of sea-weeds, after the soda has been separated by crystallisation; and it is disengaged by means of sulphuric acid, which expels it from the alkaline ley in the form of a violet gas, which may be collected and condensed in the way you have just seen. — This interesting discovery was made in the year 1812, by M. Courtois, a manufacturer of saltpetre at Paris.

CAROLINE.

And pray, Mrs. B., what is the proof of iodine being a simple body?

MRS. B.

It is considered as a simple body, both because it is not capable of being resolved into other ingredients; and because it is itself capable of combining with other bodies, in a manner analogous to oxygen and chlorine. The most curious of these combinations is that which it forms with hydrogen gas, the result of which is a peculiar gaseous acid.

CAROLINE.

Just as chlorine and hydrogen gas form muriatic acid? In this respect chlorine and iodine seem to bear a strong analogy to each other.

MRS. B.

That is indeed the case; so that if the theory of the constitution of either of these two bodies be true, it must be true also in regard to the other; if erroneous in the one, the theory must fall in both.

But it is now time to conclude; we have examined such of the acids and salts as I conceived would appear to you most interesting. — I shall not enter into any particulars respecting the metallic acids, as they offer nothing sufficiently striking for our present purpose.

CONVERSATION XX.

ON THE NATURE AND COMPOSITION OF VEGETABLES.

MRS. B.

WE have hitherto treated only of the simplest combinations of elements, such as alkalies, earths, acids, compound salts, stones, &c.; all of which belong to the mineral kingdom. It is time now to turn our attention to a more complicated class of compounds, that of ORGANISED BODIES, which will furnish us with a new source of instruction and amusement.

EMILY.

By organised bodies, I suppose, you mean the vegetable and animal creation? I have, however, but a very vague idea of the word *organisation*, and I have often wished to know more precisely what it means.

MRS. B.

Organised bodies are such as are endowed by nature with various parts, peculiarly constructed and adapted to perform certain functions connected with life. Thus you may observe, that mineral compounds are formed by the simple effect of mechanical or chemical attraction, and may appear to some to be in a great measure the productions of chance; whilst organised bodies bear the most striking and impressive marks of design, and are eminently distinguished by that unknown principle, called *life*, from which the various organs derive the power of exercising their respective functions.

CAROLINE.

But in what manner does life enable these organs to perform their several functions?

MRS. B.

That is a mystery which, I fear, is enveloped in such profound darkness that there is very little hope of our ever being able to unfold it. We must content ourselves with examining the effects of this principle; as for the cause, we have been able only to give it a name, without attaching any other meaning to it than the vague and unsatisfactory idea of an unknown agent.

CAROLINE.

And yet I think I can form a very clear idea of life.

MRS. B.

Pray let me hear how you would define it?

CAROLINE.

It is perhaps more easy to conceive than to express — let me consider — Is not life the power which enables both the animal and the vegetable creation to perform the various functions which nature has assigned to them?

MRS. B.

I have nothing to object to your definition; but you will allow me to observe, that you have only mentioned the effects which the unknown cause produces, without giving us any notion of the cause itself.

EMILY.

Yes, Caroline, you have told us what life *does*, but you have not told us what it *is*.

MRS. B.

We may study its operations, but we should puzzle ourselves to no purpose by attempting to form an idea of its real nature.

We shall begin with examining its effects in the

vegetable world, which constitutes the simplest class of organised bodies; these we shall find distinguished from the mineral creation, not only by their more complicated nature, but by the power which they possess within themselves, of forming new chemical arrangements of their constituent parts, by means of appropriate organs. Thus, though all vegetables are ultimately composed of hydrogen, carbon, and oxygen, (with a few other occasional ingredients,) they separate and combine these principles by their various organs, in a thousand ways, and form, with them, different kinds of juices and solid parts, which exist ready made in vegetables, and may, therefore, be considered as their immediate materials.

These are:

Sap,	*Resins,*
Mucilage,	*Gum Resins,*
Sugar,	*Balsams,*
Fecula,	*Caoutchouc,*
Gluten,	*Extractive colouring Matter,*
Fixed Oil,	*Tannin,*
Volatile Oil,	*Woody Fibre,*
Camphor,	*Vegetable Acids, &c.*

CAROLINE.

What a long list of names! I did not suppose

that a vegetable was composed of half so many ingredients.

MRS. B.

You must not imagine that every one of these materials is formed in each individual plant. I only mean to say, that they are all derived exclusively from the vegetable kingdom.

EMILY.

But does each particular part of the plant, such as the root, the bark, the stem, the seeds, the leaves, consist of one of these ingredients only, or of several of them combined together?

MRS. B.

I believe there is no part of a plant which can be said to consist solely of any one particular ingredient; a certain number of vegetable materials must always be combined for the formation of any particular part, (of a seed for instance,) and these combinations are carried on by sets of vessels, or minute organs, which select from other parts, and bring together, the several principles required for the development and growth of those particular parts which they are intended to form and to maintain.

EMILY.

And are not these combinations always regulated by the laws of chemical attraction?

MRS. B.

No doubt; the organs of plants cannot force principles to combine that have no attraction for each other; nor can they compel superior attractions to yield to those of inferior power; they probably act rather mechanically, by bringing into contact such principles, and in such proportions, as will, by their chemical combination, form the various vegetable products.

CAROLINE.

We may then consider each of these organs as a curiously constructed apparatus, adapted for the performance of a variety of chemical processes.

MRS. B.

Exactly so. As long as the plant lives and thrives, the carbon, hydrogen, and oxygen, (the chief constituents of its immediate materials,) are so balanced and connected together, that they are not susceptible of entering into other combinations; but no sooner does death take place, than this state of equilibrium is destroyed, and new combinations produced.

EMILY.

But why should death destroy it; for these principles must remain in the same proportions, and consequently, I should suppose, in the same order of attractions?

MRS. B.

You must remember, that in the vegetable, as well as in the animal kingdom, it is by the principle of *life* that the organs are enabled to act; when deprived of that agent or stimulus, their power ceases, and an order of attractions succeeds similar to that which would take place in mineral or unorganised matter.

EMILY.

It is this new order of attractions, I suppose, that destroys the organisation of the plant after death; for if the same combinations still continued to prevail, the plant would always remain in the state in which it died?

MRS. B.

And that, you know, is never the case; plants may be partially preserved for some time after death, by drying; but in the natural course of events they all return to the state of simple elements; a wise and admirable dispensation of Providence, by which dead plants are rendered fit to enrich the soil, and become subservient to the nourishment of living vegetables.

CAROLINE.

But we are talking of the dissolution of plants, before we have examined them in their living state.

MRS. B.

That is true, my dear. But I wished to give you a general idea of the nature of vegetation, before we entered into particulars. Besides, it is not so irrelevant as you suppose to talk of vegetables in their dead state, since we cannot analyse them without destroying life; and it is only by hastening to submit them to examination, immediately after they have ceased to live, that we can anticipate their natural decomposition. There are two kinds of analysis of which vegetables are susceptible; first, that which separates them into their immediate materials, such as sap, resin, mucilage, &c.; secondly, that which decomposes them into their primitive elements, as carbon, hydrogen, and oxygen.

EMILY.

Is there not a third kind of analysis of plants, which consists in separating their various parts, as the stem, the leaves, and the several organs of the flower?

MRS. B.

That, my dear, is rather the department of the botanist; we shall consider these different parts of plants only, as the organs by which the various secretions or separations are performed; but we must first examine the nature of these secretions.

The *sap* is the principal material of vegetables, since it contains the ingredients that nourish every part of the plant. The basis of this juice, which the roots suck up from the soil, is water; this holds in solution the various other ingredients required by the several parts of the plant, which are gradually secreted from the sap by the different organs appropriated to that purpose, as it passes them in circulating through the plant.

Mucus, or *mucilage*, is a vegetable substance, which, like all the others, is secreted from the sap; when in excess, it exudes from trees in the form of gum.

CAROLINE.

Is that the gum so frequently used instead of paste or glue?

MRS. B.

It is; almost all fruit-trees yield some sort of gum, but that most commonly used in the arts is obtained from a species of acacia-tree in Arabia, and is called *gum arabic;* it forms the chief nourishment of the natives of those parts, who obtain it in great quantities from incisions which they make in the trees.

CAROLINE.

I did not know that gum was eatable.

MRS. B.

There is an account of a whole ship's company being saved from starving by feeding on the cargo, which was gum senegal. I should not, however, imagine, that it would be either a pleasant or a particularly eligible diet to those who have not, from their birth, been accustomed to it. It is, however, frequently taken medicinally, and considered as very nourishing. Several kinds of vegetable acids may be obtained, by particular processes, from gum or mucilage, the principal of which is called the *mucous acid.*

Sugar is not found in its simple state in plants, but is always mixed with gum, sap, or other ingredients; this saccharine matter is to be met with in every vegetable, but abounds most in roots, fruits, and particularly in the sugar-cane.

EMILY.

If all vegetables contain sugar, why is it extracted exclusively from the sugar-cane?

MRS. B.

Because it is both most abundant in that plant, and most easily obtained from it. Besides, the sugars produced by other vegetables differ a little in their nature.

During the late troubles in the West-Indies, when Europe was but imperfectly supplied with sugar, several attempts were made to extract it from other vegetables, and very good sugar was obtained from parsnips and from carrots; but the process was too expensive to carry this enterprize to any extent.

CAROLINE.

I should think that sugar might be more easily obtained from sweet fruits, such as figs, dates, &c.

MRS. B.

Probably; but it would be still more expensive, from the high price of those fruits.

EMILY.

Pray, in what manner is sugar obtained from the sugar-cane?

MRS. B.

The juice of this plant is first expressed by passing it between two cylinders of iron. It is then boiled with lime-water, which makes a thick scum rise to the surface. The clarified liquor is let off below and evaporated to a very small quantity, after which it is suffered to crystallise by standing in a vessel, the bottom of which is perforated with holes, that are imperfectly stopped, in order that the syrup may drain off. The sugar obtained by this process is a coarse brown powder, commonly called raw or moist sugar; it undergoes another operation to be refined and converted into loaf sugar. For this purpose it is dissolved in water, and afterwards purified by an animal fluid called albumen. White of eggs chiefly consist of this fluid, which is also one of the constituent parts of blood; and consequently eggs, or bullocks' blood, are commonly used for this purpose.

The albuminous fluid being diffused through the syrup, combines with all the solid impurities contained in it, and rises with them to the surface, where it forms a thick scum; the clear liquor is then again evaporated to a proper consistence, and poured into moulds, in which, by a confused crystallisation, it forms loaf-sugar. But an additional process is required to whiten it; to this effect the mould is inverted, and its open base is covered with clay, through which water is made to pass; the water slowly trickling through the sugar, combines with and carries off the colouring matter.

CAROLINE.

I am very glad to hear that the blood that is used to purify sugar does not remain in it; it would be a disgusting idea. I have heard of some improvements by the late Mr. Howard, in the process of refining sugar. Pray what are they?

MRS. B.

It would be much too long to give you an account of the process in detail. But the principal improvement relates to the mode of evaporating the syrup, in order to bring it to the consistency of sugar. Instead of boiling the syrup in a large copper, over a strong fire, Mr. Howard carries off the water by means of a large air-pump, in a way similar to that used in Mr. Leslie's experiment for freezing water by evaporation; that is, the syrup being exposed to a vacuum, the water evaporates quickly, with no greater heat than that of a little steam, which is introduced round the boiler. The air-pump is of course of large dimensions, and is worked by a steam engine. A great saving is thus obtained, and a striking instance afforded of the power of science in suggesting useful economical improvements.

EMILY.

And pray how is sugar-candy and barley-sugar prepared?

MRS. B.

Candied sugar is nothing more than the regular crystals, obtained by slow evaporation from a solution of sugar. Barley-sugar is sugar melted by heat, and afterwards cooled in moulds of a spiral form.

Sugar may be decomposed by a red heat, and, like all other vegetable substances, resolved into carbonic acid and hydrogen. The formation and the decomposition of sugar afford many very interesting particulars, which we shall fully examine, after having gone through the other materials of vegetables. We shall find that there is reason to suppose that sugar is not, like the other materials, secreted from the sap by appropriate organs; but that it is formed by a peculiar process with which you are not yet acquainted.

CAROLINE.

Pray, is not honey of the same nature as sugar?

MRS. B.

Honey is a mixture of saccharine matter and gum.

EMILY.

I thought that honey was in some measure an animal substance, as it is prepared by the bees.

MRS. B.

It is rather collected by them from flowers, and conveyed to their store-houses, the hives. It is the wax only that undergoes a real alteration in the body of the bee, and is thence converted into an animal substance.

Manna is another kind of sugar, which is united with a nauseous extractive matter, to which it owes its peculiar taste and colour. It exudes like gum from various trees in hot climates, some of which have their leaves glazed by it.

The next of the vegetable materials is *fecula*; this is the general name given to the farinaceous substance contained in all seeds, and in some roots, as the potatoe, parsnip, &c. It is intended by nature for the first aliment of the young vegetable; but that of one particular grain is become a favourite and most common food of a large part of mankind.

EMILY.

You allude, I suppose, to bread, which is made of wheat-flower?

MRS. B.

Yes. The fecula of wheat contains also another vegetable substance which seems peculiar to that seed, or at least has not as yet been obtained from any other. This is *gluten*, which is of a sticky, ropy, elastic nature; and it is supposed to be owing to the viscous qualities of this substance, that wheat-flour forms a much better paste than any other.

EMILY.

Gluten, by your description, must be very like gum?

MRS. B.

In their sticky nature they certainly have some resemblance; but gluten is essentially different from gum in other points, and especially in its being insoluble in water, whilst gum, you know, is extremely soluble.

The *oils* contained in vegetables all consist of hydrogen and carbon in various proportions. They are of two kinds, *fixed* and *volatile*, both of which we formerly mentioned. Do you remember in what the difference between fixed and volatile oil consists?

EMILY.

If I recollect rightly, the former are decomposed by heat, whilst the latter are merely volatilised by it.

MRS. B.

Very well. Fixed oil is contained only in the seeds of plants, excepting in the olive, in which it is produced in, and expressed from, the fruit. We have already observed that seeds contain also fecula; these two substances, united with a little mucilage, form the white substance contained in the seeds or kernels of plants, and is destined for the nourishment of the young plant, to which the seed gives birth. The milk of almonds, which is expressed from the seed of that name, is composed of these three substances.

EMILY.

Pray, of what nature is the linseed oil which is used in painting?

MRS. B.

It is a fixed oil, obtained from the seed of flax. Nut oil, which is frequently used for the same purpose, is expressed from walnuts.

Olive oil is that which is best adapted to culinary purposes.

CAROLINE.

And what are the oils used for burning?

MRS. B.

Animal oils most commonly; but the preference given to them is owing to their being less expensive; for vegetable oils burn equally well, and are more pleasant, as their smell is not offensive.

EMILY.

Since oil is so good a combustible, what is the reason that lamps so frequently require trimming?

MRS. B.

This sometimes proceeds from the construction of the lamp, which may not be sufficiently favourable to a perfect combustion; but there is certainly a defect in the nature of oil itself, which renders it necessary for the best-constructed lamps to be occasionally trimmed. This defect arises from a portion of mucilage which it is extremely difficult to separate from the oil, and which being a bad combustible, gathers round the wick, and thus impedes its combustion, and consequently dims the light.

CAROLINE.

But will not oils burn without a wick?

MRS. B.

Not unless their temperature be elevated to five or six hundred degrees; the wick answers this purpose, as I think I once before explained to you. The oil rises between the fibres of the cotton by capillary attraction, and the heat of the burning wick volatilises it, and brings it successively to the temperature at which it is combustible.

EMILY.

I suppose the explanation which you have given with regard to the necessity of trimming lamps, applies also to candles, which so often require snuffing?

MRS. B.

I believe it does; at least, in some degree. But besides the circumstance just explained, the common sorts of oils are not very highly combustible, so that the heat produced by a candle, which is a coarse kind of animal oil, being insufficient to volatilise them completely, a quantity of soot is gradually deposited on the wick, which dims the light, and retards the combustion.

CAROLINE.

Wax candles then contain no incombustible matter, since they do not require snuffing?

MRS. B.

Wax is a much better combustible than tallow, but still not perfectly so, since it likewise contains some particles that are unfit for burning; but when these gather round the wick, (which in a wax light is comparatively small,) they weigh it down on one side, and fall off together with the burnt part of the wick.

CAROLINE.

As oils are such good combustibles, I wonder that they should require so great an elevation of temperature before they begin to burn?

MRS. B.

Though fixed oils will not enter into actual combustion below the temperature of about four hundred degrees, yet they will slowly absorb oxygen at the common temperature of the atmosphere. Hence arises a variety of changes in oils which modify their properties and uses in the arts.

If oil simply absorbs, and combines with oxygen, it thickens and changes to a kind of wax. This change is observed to take place on the external parts of certain vegetables, even during their life. But it happens in many instances that the oil does not retain all the oxygen which it attracts, but that part of it combines with, or burns, the hydrogen of the oil, thus forming a quantity of water, which gradually goes off by evaporation. In this case the alteration of the oil consists not only in the addition of a certain quantity of oxygen, but in the diminution of the hydrogen. These oils are distinguished by the name of *drying oils*. Linseed, poppy, and nut-oils, are of this description.

EMILY.

I am well acquainted with drying oils, as I continually use them in painting. But I do not understand why the acquisition of oxygen on one hand, and a loss of hydrogen on the other, should render them drying?

MRS. B.

This, I conceive, may arise from two reasons; either from the oxygen which is added being less favourable to the state of fluidity than the hydrogen, which is subtracted; or from this additional quantity of oxygen giving rise to new combinations, in con equence of which the most fluid parts of the oil are liberated and volatilised.

For the purpose of painting, the drying quality of oil is further increased by adding a quantity of oxyd of lead to it, by which means it is more rapidly oxygenated.

The rancidity of oil is likewise owing to their oxygenation. In this case a new order of attraction takes place, from which a peculiar acid is formed, called the *sebacic acid.*

CAROLINE.

Since the nature and composition of oil is so well known, pray could not oil be actually *made*, by combining its principles?

MRS. B.

That is by no means a necessary consequence; for there are innumerable varieties of compound bodies which we can decompose, although we are unable to reunite their ingredients. This, however, is not the case with oil, as it has very lately been discovered, that it is possible to form oil, by a peculiar process, from the action of oxygenated muriatic acid gas on hydro-carbonate

We now pass to the *volàtile* or *essential oils.* These form the basis of all the vegetable perfumes, and are contained, more or less, in every part of the plant excepting the seed; they are, at least, never found in that part of the seed which contains the embrio plant.

EMILY.

The smell of flowers, then, proceeds from volatile oil?

MRS. B.

Certainly; but this oil is often most abundant in the rind of fruits, as in oranges, lemons, &c. from which it may be extracted by the slightest pressure; it is found also in the leaves of plants, and even in the wood.

CAROLINE.

Is it not very plentiful in the leaves of mint, and of thyme, and all the sweet-smelling herbs?

MRS. B.

Yes, remarkably so; and in geranium leaves also, which have a much more powerful odour than the flowers.

The perfume of sandal fans is an instance of its existence in wood. In short, all vegetable odours or perfumes are produced by the evaporation of particles of these volatile oils.

EMILY.

They are, I suppose, very light, and of very thin consistence, since they are so volatile?

MRS. B.

They vary very much in this respect, some of them being as thick as butter, whilst others are as fluid as water. In order to be prepared for perfumes, or essences, these oils are first properly purified, and then either distilled with spirit of wine, as in the case with lavender water, or simply mixed with a large proportion of water, as is often done with regard to peppermint. Frequently, also, these odoriferous waters are prepared merely by soaking the plants in water, and distilling. The water then comes over impregnated with the volatile oil.

CAROLINE.

Such waters are frequently used to take spots of grease out of cloth, or silk; how do they produce that effect?

MRS. B.

By combining with the substance that forms these stains; for volatile oils, and likewise the spirit in which they are distilled, will dissolve wax, tallow, spermaceti, and resins; if, therefore, the spot proceeds from any of these substanoes, it will remove it. Insects of every kind have a great aversion to perfumes, so that volatile oils are employed with success in museums for the preservation of stuffed birds and other species of animals.

CAROLINE.

Pray does not the powerful smell of camphor proceed from a volatile oil?

MRS. B.

Camphor seems to be a substance of its own kind, remarkable by many peculiarities. But if not exactly of the same nature as volatile oil, it is at least very analogous to it. It is obtained chiefly from the camphor-tree, a species of laurel which grows in China, and in the Indian isles, from the stem and roots of which it is extracted. Small quantities have also been distilled from thyme, sage, and other aromatic plants; and it is deposited in pretty large quantities by some volatile oils after long standing. It is extremely volatile and inflammable. It is insoluble in water, but is soluble in oils, in which state, as well as in its solid form, it is frequently applied to medicinal purposes. Amongst the particular properties of camphor, there is one too singular to be passed over in silence. If you take a small piece of camphor, and place it on the surface of a bason of pure water, it will immediately begin to move round and round with great rapidity; but if you pour into the bason a single drop of any odoriferous fluid, it will instantly put a stop to this motion. You can at any time try this very simple experiment; but you must not expect that I shall be able to account for this phenomenon, as nothing satisfactory has yet been advanced for its explanation.

CAROLINE.

It is very singular indeed; and I will certainly try the experiment. Pray what are *resins*, which you just now mentioned?

MRS. B.

They are volatile oils, that have been acted on, and peculiarly modified, by oxygen.

CAROLINE.

They are, therefore, oxygenated volatile oils?

MRS. B.

Not exactly; for the process does not appear to consist so much in the oxygenation of the oil, as in the combustion of a portion of its hydrogen, and a small portion of its carbon. For when resins are artificially made by the combination of volatile oils with oxygen, the vessel in which the process is performed is bedewed with water, and the air included within is loaded with carbonic acid.

EMILY.

This process must be, in some respects, similar to that for preparing drying oils?

MRS. B.

Yes; and it is by this operation that both of them acquire a greater degree of consistence. Pitch, tar, and turpentine, are the most common resins; they exude from the pine and fir trees. Copal, mastic, and frankincense, are also of this class of vegetable substances.

EMILY.

Is it of these resins that the mastic and copal varnishes, so much used in painting, are made?

MRS. B.

Yes. Dissolved either in oil, or in alcohol, resins form varnishes. From these solutions they may be precipitated by water, in which they are insoluble. This I can easily show you. — If you will pour some water into this glass of mastic varnish, it will combine with the alcohol in which the resin is dissolved, and the latter will be precipitated in the form of a white cloud —

EMILY.

It is so. And yet how is it that pictures or

drawings, varnished with this solution, may safely be washed with water?

MRS. B.

As the varnish dries, the alcohol evaporates, and the dry varnish or resin which remains, not being soluble in water, will not be acted on by it.

There is a class of compound resins called *gum-resins*, which are precisely what their name denotes, that is to say, resins combined with mucilage. Myrrh and assafœtida are of this description.

CAROLINE.

Is it possible that a substance of so disagreeable a smell as assafœtida can be formed from a volatile oil?

MRS. B.

The odour of volatile oils is by no means always grateful. Onions and garlic derive their smell from volatile oils, as well as roses and lavender.

There is still another form under which volatile oils present themselves, which is that of *balsams.* These consist of resinous juices combined with a peculiar acid, called the benzoic acid. Balsams appear to have been originally volatile oils, the oxygenation of which has converted one part into a resin, and the other part into an acid, which, combined together, form a balsam; such are the balsams of Peru, Tolu, &c.

We shall now take leave of the oils and their various modifications, and proceed to the next vegetable substance, which is *caoutchouc*. This is a white milky glutinous fluid, which acquires consistence, and blackens in drying, in which state it forms the substance with which you are so well acquainted, under the name of gum-elastic.

CAROLINE.

I am surprised to hear that gum-elastic was ever white, or ever fluid! And from what vegetable is it procured?

MRS. B.

It is obtained from two or three different species of trees, in the East-Indies, and South-America, by making incisions in the stem. The juice is collected as it trickles from these incisions, and moulds of clay, in the form of little bottles of gum-elastic, are dipped into it. A layer of this juice adheres to the clay and dries on it; and several layers are successively added by repeating this till the bottle is of sufficient thickness. It is then beaten to break down the clay, which is easily shaken out. The natives of the countries where this substance is produced sometimes make shoes and boots of it by a similar process, and they are said to be extremely pleasant and serviceable, both from their elasticity, and their being water-proof.

The substance which comes next in our enumeration of the immediate ingredients of vegetables, is *extractive matter*. This is a term, which, in a general sense, may be applied to any substance extracted from vegetables; but it is more particularly understood to relate to the extractive *colouring matter* of plants. A great variety of colours are prepared from the vegetable kingdom, both for the purposes of painting and of dying; all the colours called *lakes* are of this description; but they are less durable than mineral colours, for, by long exposure to the atmosphere, they either darken or turn yellow.

EMILY.

I know that in painting, the lakes are reckoned far less durable colours than the ochres; but what is the reason of it?

MRS. B.

The change which takes place in vegetable colours is owing chiefly to the oxygen of the atmosphere slowly burning their hydrogen, and leaving, in some measure, the blackness of the carbon exposed. Such change cannot take place in ochre, which is altogether a mineral substance.

Vegetable colours have a stronger affinity for animal than for vegetable substances, and this is supposed to be owing to a small quantity of nitrogen which they contain. Thus, silk and worsted will take a much finer vegetable dye than linen and cotton.

CAROLINE.

Dying, then, is quite a chemical process?

MRS. B.

Undoubtedly. The condition required to form a good dye is, that the colouring matter should be precipitated, or fixed, on the substance to be dyed, and should form a compound not soluble in the liquids to which it will probably be exposed. Thus, for instance, printed or dyed linens or cottons must be able to resist the action of soap and water, to which they must necessarily be subject in washing; and woollens and silks should withstand the action of grease and acids, to which they may accidentally be exposed.

CAROLINE.

But if linen and cotton have not a sufficient affinity for colouring matter, how are they made to resist the action of washing, which they always do when they are well printed?

MRS. B.

When the substance to be dyed has either no affinity for the colouring matter, or not sufficient power to retain it, the combination is effected, or strengthened, by the intervention of a third substance, called a *mordant*, or basis. The mordant must have a strong affinity both for the colouring matter and the substance to be dyed, by which means it causes them to combine and adhere together.

CAROLINE.

And what are the substances that perform the office of thus reconciling the two adverse parties?

MRS. B.

The most common mordant is sulphat of alumine, or alum. Oxyds of tin and iron, in the state of compound salts, are likewise used for that purpose.

Tannin is another vegetable ingredient of great importance in the arts. It is obtained chiefly from the bark of trees; but it is found also in nut-galls, and in some other vegetables.

EMILY.

Is that the substance commonly called *tan*, which is used in hot-houses?

MRS. B.

Tan is the prepared bark in which the peculiar substance, tannin, is contained. But the use of tan in hot-houses is of much less importance than in the operation of *tanning*, by which skin is converted into leather.

EMILY.

Pray, how is this operation performed?

MRS. B.

Various methods are employed for this purpose, which all consist in exposing skin to the action of tannin, or of substances containing this principle, in sufficient quantities, and disposed to yield it to the skin. The most usual way is to infuse coarsely powdered oak bark in water, and to keep the skin immersed in this infusion for a certain length of time. During this process, which is slow and gradual, the skin is found to have increased in weight, and to have acquired a considerable tenacity and impermeability to water. This effect may be much accelerated by using strong saturations of the tanning principle (which can be extracted from bark), instead of employing the bark itself. But this quick mode of preparation does not appear to make equally good leather.

Tannin is contained in a great variety of

astringent vegetable substances, as galls, the rose-tree, and wine; but it is nowhere so plentiful as in bark. All these substances yield it to water, from which it may be precipitated by a solution of isinglass, or glue, with which it strongly unites and forms an insoluble compound. Hence its valuable property of combining with skin (which consists chiefly of glue), and of enabling it to resist the action of water.

EMILY.

Might we not see that effect by pouring a little melted isinglass into a glass of wine, which you say contains tannin?

MRS. B.

Yes. I have prepared a solution of isinglass for that very purpose. — Do you observe the thick muddy precipitate? — That is the tannin combined with the isinglass.

CAROLINE.

This precipitate must then be of the same nature as leather?

MRS. B.

It is composed of the same ingredients; but the organisation and texture of the skin being wanting, it has neither the consistence nor the tenacity of leather.

CAROLINE.

One might suppose that men who drink large quantities of red wine stand a chance of having the coats of their stomachs converted into leather, since tannin has so strong an affinity for skin.

MRS. B.

It is not impossible but that the coats of their stomachs may be, in some measure, tanned, or hardened by the constant use of this liquor; but you must remember that where a number of other chemical agents are concerned, and, above all, where life exists, no certain chemical inference can be drawn.

I must not dismiss this subject, without mentioning a recent discovery of Mr. Hatchett, which relates to it. This gentleman found that a substance very similar to tannin, possessing all its leading properties, and actually capable of tanning leather, may be produced by exposing carbon, or any substance containing carbonaceous matter, whether vegetable, animal, or mineral, to the action of nitric acid.

CAROLINE.

And is not this discovery very likely to be of use to manufactures?

MRS. B.

That is very doubtful, because tannin, thus artificially prepared, must probably always be more expensive than that which is obtained from bark. But the fact is extremely curious, as it affords one of those very rare instances of chemistry being able to imitate the proximate principles of organised bodies.

The last of the vegetable materials is *woody fibre;* it is the hardest part of plants. The chief source from which this substance is derived is wood, but it is also contained, more or less, in every solid part of that plant. It forms a kind of skeleton of the part to which it belongs, and retains its shape after all the other materials have disappeared. It consists chiefly of carbon, united with a small proportion of salts, and the other constituents common to all vegetables.

EMILY.

It is of woody fibre, then, that the common charcoal is made?

MRS. B.

Yes. Charcoal, as you may recollect, is obtained from wood, by the separation of all its evaporable parts.

Before we take leave of the vegetable materials, it will be proper, at least, to enumerate the several vegetable acids which we either have had, or may have occasion to mention. I believe I formerly told you that their basis, or radical, was uniformly composed of hydrogen and carbon, and that their difference consisted only in the various proportions of oxygen which they contained.

The following are the names of the vegetable acids:

The *Mucous Acid*, obtained from gum or mucilage;

Suberic - - -	from cork;
Camphoric - -	from camphor;
Benzoic - -	from balsams;
Gallic - - - -	from galls, bark, &c.
Malic - - -	from ripe fruits;
Citric - - -	from lemon juice;
Oxalic - - -	from sorrel;
Succinic - - -	from amber;
Tartarous - - -	from tartrit of potash;
Acetic - - -	from vinegar.

They are all decomposable by heat, soluble in water, and turn vegetable blue colours red. The *succinic*, the *tartarous*, and the *acetous acids*, are the products of the decomposition of vegetables, we shall, therefore, reserve their examination for a future period.

The *oxalic acid*, distilled from sorrel, is the highest term of vegetable acidification; for, if more oxygen be added to it, it loses its vegetable nature, and is resolved into carbonic acid and water; therefore, though all the other acids may be converted into the oxalic by an addition of oxygen, the oxalic itself is not susceptible of a further degree of oxygenation; nor can it be made, by any chemical processes, to return to a state of lower acidification.

To conclude this subject, I have only to add a few words on the *gallic acid*. . . . .

CAROLINE.

Is not this the same acid before mentioned, which forms ink, by precipitating sulphat of iron from its solution?

MRS. B.

Yes. Though it is usually extracted from galls, on account of its being most abundant in that vegetable substance, it may also be obtained from a great variety of plants. It constitutes what is called the *astringent principle* of vegetables; it is generally combined with tannin, and you will find that an infusion of tea, coffee, bark, red-wine, or any vegetable substance that contains the astringent principle, will make a black precipitate with a solution of sulphat of iron.

CAROLINE.

But pray what are galls?

MRS. B.

They are excrescences which grow on the bark of young oaks, and are occasioned by an insect which wounds the bark of trees, and lays its eggs in the aperture. The lacerated vessels of the tree then discharge their contents, and form an excrescence, which affords a defensive covering for these eggs. The insect, when come to life, first feeds on this excrescence, and some time afterward eats its way out, as it appears from a hole which is formed in all gall-nuts that no longer contain an insect. It is in hot climates only that strongly astringent gall-nuts are found; those which are used for the purpose of making ink are brought from Aleppo.

EMILY.

But are not the oak-apples, which grow on the leaves of the oak in this country, of a similar nature?

MRS. B.

Yes; only they are an inferior species of galls, containing less of the astringent principle, and therefore less applicable to useful purposes.

CAROLINE.

Are the vegetable acids never found but in their pure uncombined state?

MRS. B.

By no means; on the contrary, they are frequently met with in the state of compound salts; these, however, are in general not fully saturated with the salifiable bases, so that the acid predominates; and, in this state, they are called *acidulous* salts. Of this kind is the salt called cream of tartar.

CAROLINE.

Is not the salt of lemon, commonly used to take out ink-spots and stains, of this nature?

MRS. B.

No; that salt consists of the oxalic acid, combined with a little potash. It is found in that state in sorrel.

CAROLINE.

And pray how does it take out ink-spots?

MRS. B.

By uniting with the iron, and rendering it soluble in water.

Besides the vegetable materials which we have enumerated, a variety of other substances, common to the three kingdoms, are found in vegetables, such as potash, which was formerly supposed to belong exclusively to plants, and was, in consequence, called the vegetable alkali.

Sulphur, phosphorus, earths, and a variety of metallic oxyds, are also found in vegetables, but only in small quantities. And we meet sometimes with neutral salts, formed by the combination of these ingredients.

CONVERSATION XXI.

ON THE DECOMPOSITION OF VEGETABLES.

CAROLINE.

THE account which you have given us, Mrs. B., of the materials of vegetables, is, doubtless, very instructive; but it does not completely satisfy my curiosity. I wish to know how plants obtain the principles from which their various materials are formed; by what means these are converted into vegetable matter, and how they are connected with the life of the plant?

MRS. B.

This implies nothing less than a complete history of the chemistry and physiology of vegetation, subjects on which we have yet but very imperfect notions. Still I hope that I shall be able, in some measure, to satisfy your curiosity. But, in order to render the subject more intelligible, I must first make you acquainted with the various changes which vegetables undergo, when the vital power no longer enables them to resist the common laws of chemical attraction.

The composition of vegetables being more complicated than that of minerals, the former more readily undergo chemical changes than the latter: for the greater the variety of attractions, the more easily is the equilibrium destroyed, and a new order of combinations introduced.

EMILY.

I am surprised that vegetables should be so easily susceptible of decomposition; for the preservation of the vegetable kingdom is certainly far more important than that of minerals.

MRS. B.

You must consider, on the other hand, how much more easily the former is renewed than the latter. The decomposition of the vegetable takes place only after the death of the plant, which, in the common course of nature, happens when it has yielded fruit and seeds to propagate its species. If, instead of thus finishing its career, each plant was to retain its form and vegetable state, it would become an useless burden to the earth and its inhabitants. When vegetables, therefore, cease to be productive, they cease to live, and nature then begins her process of decomposition, in order to resolve them into their chemical constituents, hydrogen, carbon, and oxygen; those simple and primitive ingredients, which she keeps in store for all her combinations.

EMILY.

But since no system of combination can be destroyed, except by the establishment of another order of attractions, how can the decomposition of vegetables reduce them to their simple elements?

MRS. B.

It is a very long process, during which a variety of new combinations are successively established and successively destroyed: but, in each of these changes, the ingredients of vegetable matter tend to unite in a more simple order of compounds, till they are at length brought to their elementary state, or, at least, to their most simple order of combinations. Thus you will find that vegetables are in the end almost entirely reduced to water and carbonic acid; the hydrogen and carbon dividing the oxygen between them, so as to form with it these two substances. But the variety of intermediate combinations that take place during the several stages of the decomposition of vegetables, present us with a new set of compounds, well worthy of our examination.

CAROLINE.

How is it possible that vegetables, while putrefying, should produce any thing worthy of observation?

MRS. B.

They are susceptible of undergoing certain changes before they arrive at the state of putrefaction, which is the final term of decomposition; and of these changes we avail ourselves for particular and important purposes. But, in order to make you understand this subject, which is of considerable importance, I must explain it more in detail.

The decomposition of vegetables is always attended by a violent internal motion, produced by the disunion of one order of particles, and the combination of another. This is called FERMENTATION. There are several periods at which this process stops, so that a state of rest appears to be restored, and the new order of compounds fairly established. But, unless means be used to secure these new combinations in their actual state, their duration will be but transient, and a new fermentation will take place, by which the compound last formed will be detroyed; and another, and less complex order, will succeed.

EMILY.

The fermentations, then, appear to be only the

successive steps by which a vegetable descends to its final dissolution.

MRS. B.

Precisely so. Your definition is perfectly correct.

CAROLINE.

And how many fermentations, or new arrangements, does a vegetable undergo before it is reduced to its simple ingredients?

MRS. B.

Chemists do not exactly agree in this point; but there are, I think, four distinct fermentations, or periods, at which the decomposition of vegetable matter stops and changes its course. But every kind of vegetable matter is not equally susceptible of undergoing all these fermentations.

There are likewise several circumstances required to produce fermentation. Water and a certain degree of heat are both essential to this process, in order to separate the particles, and thus weaken their force of cohesion, that the new chemical affinities may be brought into action.

CAROLINE.

In frozen climates, then, how can the spontaneous decomposition of vegetables take place?

MRS. B.

It certainly cannot; and, accordingly, we find scarcely any vestiges of vegetation where a constant frost prevails.

CAROLINE.

One would imagine that, on the contrary, such spots would be covered with vegetables; for, since they cannot be decomposed, their number must always increase.

MRS. B.

But, my dear, heat and water are quite as essential to the formation of vegetables, as they are to their decomposition. Besides, it is from the dead vegetables, reduced to their elementary principles, that the rising generation is supplied with sustenance. No young plant, therefore, can grow unless its predecessors contribute both to its formation and support; and these not only furnish the seed from which the new plant springs, but likewise the food by which it is nourished.

CAROLINE.

Under the torrid zone, therefore, where water is never frozen, and the heat is very great, both the processes of vegetation and of fermentation must, I suppose, be extremely rapid?

MRS. B.

Not so much as you imagine: for in such climates great part of the water which it requires for these processes is in an aëriform state, which is scarcely more conducive either to the growth or formation of vegetables than that of ice. In those latitudes, therefore, it is only in low damp situations, sheltered by woods from the sun's rays, that the smaller tribes of vegetables can grow and thrive during the dry season, as dead vegetables seldom retain water enough to produce fermentation, but are, on the contrary, soon dried up by the heat of the sun, which enables them to resist that process; so that it is not till the fall of the autumnal rains (which are very violent in such climates), that spontaneous fermentation can take place.

The several fermentations derive their names from their principal products. The first is called the *saccharine fermentation*, because its product is *sugar*.

CAROLINE.

But sugar, you have told us, is found in all vegetables; it cannot, therefore, be the product of their decomposition.

MRS. B.

It is true that this fermentation is not confined to the decomposition of vegetables, as it continually takes place during their life; and, indeed, this circumstance has, till lately, prevented it from being considered as one of the fermentations. But the process appears so analogous to the other fermentations, and the formation of sugar, whether in living or dead vegetable matter is so evidently a new compound, proceeding from the destruction of the previous order of combinations, and essential to the subsequent fermentations, that it is now, I believe, generally esteemed the first step, or necessary preliminary, to decomposition, if not an actual commencement of that process.

CAROLINE.

I recollect your hinting to us that sugar was supposed not to be secreted from the sap, in the same manner as mucilage, fecula, oil, and the other ingredients of vegetables.

MRS. B.

It is rather from these materials, than from the sap itself, that sugar is formed; and it is developed at particular periods, as you may observe in fruits, which become sweet in ripening, sometimes even after they have been gathered. Life, therefore, is not essential to the formation of sugar, whilst on the contrary, mucilage, fecula, and the other vegetable materials that are secreted from the sap by appropriate organs, whose powers immediately depend on the vital principle, cannot be produced but during the existence of that principle.

EMILY.

The ripening of fruits is, then, their first step to destruction, as well as their last towards perfection?

MRS. B.

Exactly. — A process analogous to the saccharine fermentation takes place also during the cooking of certain vegetables. This is the case with parsnips, carrots, potatoes, &c. in which sweetness is developed by heat and moisture; and we know that if we carried the process a little farther, a more complete decomposition would ensue. The same process takes place also in seeds previous to their sprouting.

CAROLINE.

How do you reconcile this to your theory, Mrs. B.? Can you suppose that a decomposition is the necessary precursor of life?

MRS. B.

That is indeed the case. The materials of the seed must be decomposed, and the seed disorganized, before a plant can sprout from it. Seeds, besides the embrio plant, contain (as we have already observed) fecula, oil, and a little mucilage. These substances are destined for the nourishment of the future plant; but they undergo some change before they can be fit for this function. The seeds, when buried in the earth, with a certain degree of moisture and of temperature, absorb water, which dilates them, separates their particles, and introduces a new order of attractions, of which sugar is the product. The substance of the seed is thus softened, sweetened, and converted into a sort of white milky pulp, fit for the nourishment of the embrio plant.

The saccharine fermentation of seeds is artificially produced, for the purpose of making *malt*, by the following process: — A quantity of barley is first soaked in water for two or three days: the water being afterwards drained off, the grain heats spontaneously, swells, bursts, sweetens, shows a disposition to germinate, and actually sprouts to the length of an inch, when the process is stopped by putting it into a kiln, where it is well dried at a gentle heat. In this state it is crisp and friable, and constitutes the substance called *malt*, which is the principal ingredient of beer.

EMILY.

But I hope you will tell us how malt is made into beer?

MRS. B.

Certainly; but I must first explain to you the nature of the second fermentation, which is essential to that operation. This is called the *vinous fermentation*, because its product is *wine*.

EMILY.

How very different the decomposition of vegetables is from what I had imagined! The products of their disorganisation appear almost superior to those which they yield during their state of life and perfection.

MRS. B.

And do you not, at the same time, admire the beautiful economy of Nature, which, whether she creates, or whether she destroys, directs all her operations to some useful and benevolent purpose? — It appears that the saccharine fermentation is extremely favourable, if not absolutely essential, as a previous step, to the vinous fermentation; so that if sugar be not developed during the life of the plant, the saccharine fermentation must be artificially produced before the vinous fermentation can take place. This is the case with barley, which does not yield any sugar until it is made into malt; and it is in that state only that it is susceptible of undergoing the vinous fermentation by which it is converted into beer.

CAROLINE.

But if the product of the vinous fermentation

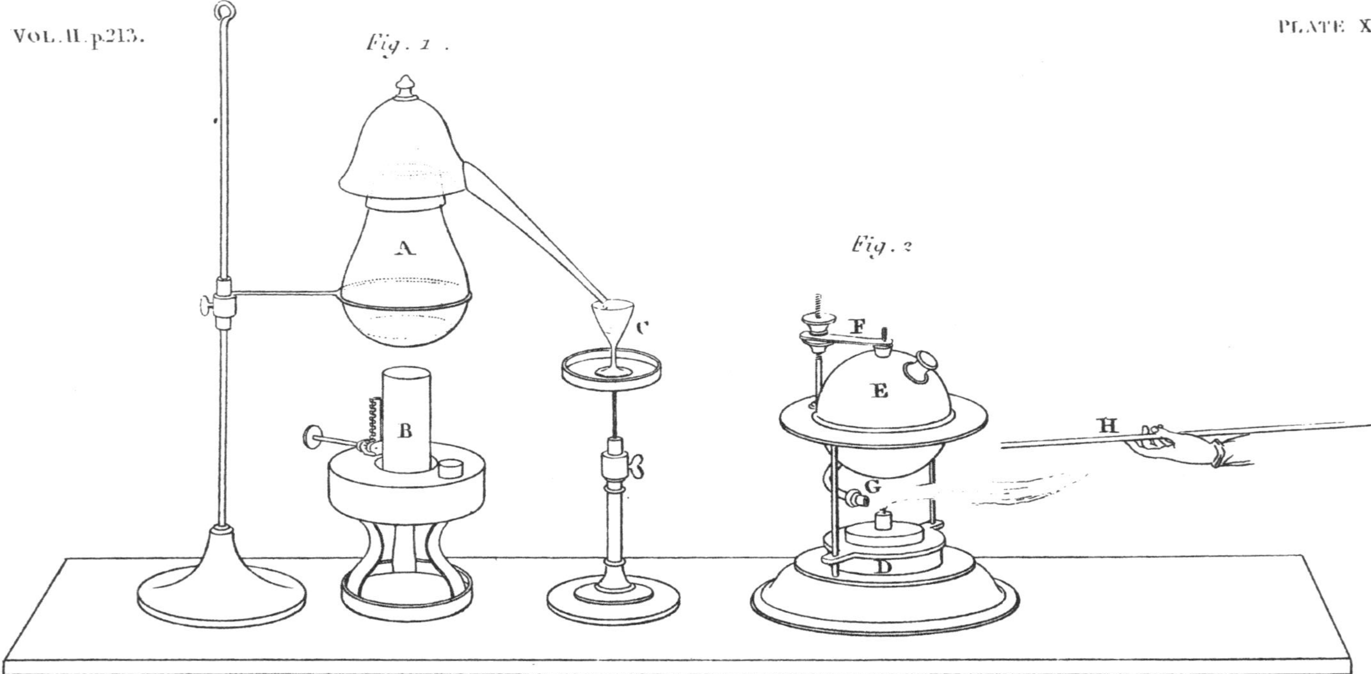

Fig. 1. A *Alembic.* _B *Lamp.* _C *Wine glass.* _*Fig. 2. Alcohol blowpipe.* _D *the Lamp.* _E *the vessel in which Alcohol is boiling.* F *a safety valve.* _G *the inflamed jet or steam of alcohol directed towards a glass tube* H.

Drawn by the Author. Published by Longman & Co. Oct.r 2nd 1809. Engraved by Lowry.

is always wine, beer cannot have undergone that process, for beer is certainly not wine.

MRS. B.

Chemically speaking, beer may be considered as the wine of grain. For it is the product of the fermentation of malt, just as wine is that of the fermentation of grapes, or other fruits.

The consequence of the vinous fermentation is the decomposition of the saccharine matter, and the formation of a spirituous liquor from the constituents of the sugar. But, in order to promote this fermentation, not only water and a certain degree of heat are necessary, but also some other vegetable ingredients, besides the sugar, as fecula, mucilage, acids, salts, extractive matter, &c. all of which seem to contribute to this process; and give to the liquor its peculiar taste.

EMILY.

It is, perhaps, for this reason that wine is not obtained from the fermentation of pure sugar; but that fruits are chosen for that purpose, as they contain not only sugar, but likewise the other vegetable ingredients which promote the vinous fermentation, and give the peculiar flavour.

MRS. B.

Certainly. And you must observe also, that

the relative quantity of sugar is not the only circumstance to be considered in the choice of vegetable juices for the formation of wine; otherwise the sugar-cane would be best adapted for that purpose. It is rather the manner and proportion in which the sugar is mixed with other vegetable ingredients that influences the production and qualities of wine. And it is found that the juice of the grape not only yields the most considerable proportion of wine, but that it likewise affords it of the most grateful flavour.

EMILY.

I have seen a vintage in Switzerland, and I do not recollect that heat was applied, or water added, to produce the fermentation of the grapes.

MRS. B.

The common temperature of the atmosphere in the cellars in which the juice of the grape is fermented is sufficiently warm for this purpose; and as the juice contains an ample supply of water, there is no occasion for any addition of it. But when fermentation is produced in dry malt, a quantity of water must necessarily be added.

EMILY.

But what are precisely the changes that happen during the vinous fermentation?

MRS. B.

The sugar is decomposed, and its constituents are recombined into two new substances; the one a peculiar liquid substance, called *alcohol* or *spirit of wine*, which remains in the fluid; the other, carbonic acid gas, which escapes during the fermentation. Wine, therefore, as I before observed, in a general point of view, may be considered as a liquid of which alcohol constitutes the essential part. And the varieties of strength and flavour of the different kinds of wine are to be attributed to the different qualities of the fruits from which they are obtained, independently of the sugar.

CAROLINE.

I am astonished to hear that so powerful a liquid as spirit of wine should be obtained from so mild a substance as sugar.

MRS. B.

Can you tell me in what the principal difference consists between alcohol and sugar?

CAROLINE.

Let me reflect . . . . . Sugar consists of carbon, hydrogen, and oxygen. If carbonic acid be subtracted from it, during the formation of alcohol, the latter will contain less carbon and oxygen than sugar does; therefore hydrogen must be the prevailing principle of alcohol.

MRS. B.

It is exactly so. And this very large proportion of hydrogen accounts for the lightness and combustible property of alcohol, and of spirits in general, all of which consist of alcohol variously modified.

EMILY.

And can sugar be recomposed from the combination of alcohol and carbonic acid?

MRS. B.

Chemists have never been able to succeed in effecting this; but from analogy, I should suppose such a recomposition possible. Let us now observe more particularly the phenomena that take place during the vinous fermentation. At the commencement of this process, heat is evolved, and the liquor swells considerably from the formation of the carbonic acid, which is disengaged in such prodigious quantities as would be fatal to any person who should unawares inspire it; an accident which has sometimes happened. If the fermentation be stopped by putting the liquor into barrels, before the whole of the carbonic acid is evolved, the wine is brisk, like Champagne, from the carbonic acid imprisoned in it, and it tastes sweet, like cyder, from the sugar not being completely decomposed.

EMILY.

But I do not understand why heat should be evolved during this operation. For, as there is a considerable formation of gas, in which a proportionable quantity of heat must become insensible, I should have imagined that cold, rather than heat, would have been produced.

MRS. B.

It appears so on first consideration; but you must recollect that fermentation is a complicated chemical process; and that, during the decompositions and recompositions attending it, a quantity of chemical heat may be disengaged, sufficient both to develope the gas, and to effect an increase of temperature. When the fermentation is completed, the liquid cools and subsides, the effervescence ceases, and the thick, sweet, sticky juice of the fruit is converted into a clear, transparent, spirituous liquor, called wine.

EMILY.

How much I regret not having been acquainted with the nature of the vinous fermentation, when I had an opportunity of seeing the process!

MRS. B.

You have an easy method of satisfying yourself in that respect by observing the process of brewing, which, in every essential circumstance, is similar to that of making wine, and is really a very curious chemical operation.

Although we cannot actually make wine at this moment, it will be easy to show you the mode of analyzing it. This is done by distillation. When wine of any kind is submitted to this operation, it is found to contain brandy, water, tartar, extractive colouring matter, and some vegetable acids. I have put a little port wine into this alembic of glass (PLATE XIV. Fig. 1.), and on placing the lamp under it, you will soon see the spirit and water successively come over—

EMILY.

But you do not mention alcohol amongst the *products* of the distillation of wine; and yet that is its most essential ingredient?

MRS. B.

The alcohol is contained in the brandy which is now coming over, and dropping from the still. Brandy is nothing more than a mixture of alcohol and water; and in order to obtain the alcohol pure, we must again distil it from brandy.

CAROLINE.

I have just taken a drop on my finger; it tastes like strong brandy, but it is without colour, whilst brandy is of a deep yellow.

MRS. B.

It is not so naturally; in its pure state brandy is colourless, and it obtains the yellow tint you observe, by extracting the colouring matter from the new oaken casks in which it is kept. But if it does not acquire the usual tinge in this way, it is the custom to colour the brandy used in this country artificially, with a little burnt sugar, in order to give it the appearance of having been long kept.

CAROLINE.

And is rum also distilled from wine?

MRS. B.

By no means; it is distilled from the sugar-cane, a plant which contains so great a quantity of sugar, that it yields more alcohol than almost any other vegetable. After the juice of the cane has been pressed out for making sugar, what still remains in the bruised cane is extracted by water, and this watery solution of sugar is fermented, and produces rum.

The spirituous liquor called *arack* is in a similar manner distilled from the product of the vinous fermentation of rice.

EMILY.

But rice has no sweetness; does it contain any sugar?

MRS. B.

Like barley and most other seeds, it is insipid until it has undergone the saccharine fermentation; and this, you must recollect, is always a previous step to the vinous fermentation in those vegetables in which sugar is not already formed. Brandy may in the same manner be obtained from malt.

CAROLINE.

You mean from beer, I suppose; for the malt must have previously undergone the vinous fermentation.

MRS. B.

Beer is not precisely the product of the vinous fermentation of malt. For hops are a necessary ingredient for the formation of that liquor; whilst brandy is distilled from pure fermented malt. But brandy might, no doubt, be distilled from beer as well as from any other liquor that has undergone the vinous fermentation; for since the basis of brandy is alcohol, it may be obtained from any liquid that contains that spirituous substance.

EMILY.

And pray, from what vegetable is the favourite spirit of the lower orders of people, gin, extracted?

MRS. B.

The spirit (which is the same in all fermented liquors) may be obtained from any kind of grain; but the peculiar flavour which distinguishes gin is that of juniper berries, which are distilled together with the grain—

I think the brandy contained in the wine which we are distilling must, by this time, be all come over. Yes—taste the liquid that is now dropping from the alembic—

CAROLINE.

It is perfectly insipid, like water.

MRS. B.

It is water, which, as I was telling you, is the second product of wine, and comes over after all the spirit, which is the lightest part, is distilled.—The tartar and extractive colouring matter we shall find in a solid form at the bottom of the alembic.

EMILY.

They look very like the lees of wine.

MRS. B.

And in many respects they are of a similar nature; for lees of wine consist chiefly of tartrit of potash; a salt which exists in the juice of the grape, and in many other vegetables, and is developed only by the vinous fermentation. During this operation it is precipitated, and deposits itself on the internal surface of the cask in which the wine is contained. It is much used in medicine, and in various arts, particularly dying, under the name of *cream of tartar*, and it is from this salt that the tartarous acid is obtained.

CAROLINE.

But the medicinal cream of tartar is in appearance quite different from these dark-coloured dregs; it is perfectly colourless.

MRS. B.

Because it consists of the pure salts only, in its crystallised form; whilst in the instance before us it is mixed with the deep-coloured extractive matter, and other foreign ingredients.

EMILY.

Pray cannot we now obtain pure alcohol from the brandy which we have distilled?

MRS. B.

We might; but the process would be tedious: for in order to obtain alcohol perfectly free from water, it is necessary to distil, or, as the distillers call it, *rectify* it several times. You must therefore allow me to produce a bottle of alcohol that has been thus purified. This is a very important ingredient, which has many striking properties, besides its forming the basis of all spirituous liquors.

EMILY.

It is alcohol, I suppose, that produces intoxication?

MRS. B.

Certainly; but the stimulus and momentary energy it gives to the system, and the intoxication it occasions when taken in excess, are circumstances not yet accounted for.

CAROLINE.

I thought that it produced these effects by increasing the rapidity of the circulation of the blood; for drinking wine or spirits, I have heard, always quickens the pulse.

MRS. B.

No doubt; the spirit, by stimulating the nerves, increases the action of the muscles; and the heart, which is one of the strongest muscular organs, beats with augmented vigour, and propels the blood with accelerated quickness. After such a strong excitation the frame naturally suffers a proportional degree of depression, so that a state of debility and langour is the invariable consequence of intoxication. But though these circumstances are well ascertained, they are far from explaining why alcohol should produce such effects.

EMILY.

Liqueurs are the only kind of spirits which I think pleasant. Pray of what do they consist?

MRS. B.

They are composed of alcohol, sweetened with syrup, and flavoured with volatile oil.

The different kinds of odoriferous spirituous waters are likewise solutions of volatile oil in alcohol, as lavender water, eau de Cologne, &c.

The chemical properties of alcohol are important and numerous. It is one of the most powerful chemical agents, and is particularly useful in dissolving a variety of substances, which are soluble neither by water nor heat.

EMILY.

We have seen it dissolve copal and mastic to form varnishes; and these resins are certainly not soluble in water, since water precipitates them from their solution in alcohol.

MRS. B.

I am happy to find that you recollect these circumstances so well. The same experiment affords also an instance of another property of alcohol, — its tendency to unite with water; for the resin is precipitated in consequence of losing the alcohol, which abandons it from its preference for water. It is attended also, as you may recollect, with the same peculiar circumstance of a disengagement of heat and consequent diminution of bulk, which we have supposed to be produced by a mechanical penetration of particles by which latent heat is forced out.

Alcohol unites thus readily not only with resins and with water, but with oils and balsams; these compounds form the extensive class of elixirs, tinctures, quintessences, &c.

EMILY.

I suppose that alcohol must be highly combustible, since it contains so large a proportion of hydrogen?

MRS. B.

Extremely so; and it will burn at a very moderate temperature.

CAROLINE.

I have often seen both brandy and spirit of

wine burnt; they produce a great deal of flame, but not a proportional quantity of heat, and no smoke whatever.

MRS. B.

The last circumstance arises from their combustion being complete; and the disproportion between the flame and heat shows you that these are by no means synonymous.

The great quantity of flame proceeds from the combustion of the hydrogen to which, you know, that manner of burning is peculiar. — Have you not remarked also that brandy and alcohol will burn without a wick? — They take fire at so low a temperature, that this assistance is not required to concentrate the heat and volatilise the fluid.

CAROLINE.

I have sometimes seen brandy burnt by merely heating it in a spoon.

MRS. B.

The rapidity of the combustion of alcohol may, however, be prodigiously increased by first volatilising it. An ingenious instrument has been constructed on this principle to answer the purpose of a blow-pipe, which may be used for melting glass, or other chemical purposes. It consists of a small metallic vessel (Plate XIII. Fig. 2.), of a spherical shape, which contains the alcohol, and is heated by the lamp beneath it; as soon as the alcohol is volatilised, it passes through the spout of the vessel, and issues just above the wick of the lamp, which immediately sets fire to the stream of vapour, as I shall show you —

EMILY.

With what amazing violence it burns! The flame of alcohol, in the state of vapour, is, I fancy, much hotter than when the spirit is merely burnt in a spoon?

MRS. B.

Yes; because in this way the combustion goes on much quicker, and, of course, the heat is proportionally increased. — Observe its effect on this small glass tube, the middle of which I present to the extremity of the flame, where the heat is greatest.

CAROLINE.

The glass, in that spot, is become red hot, and bends from its own weight.

MRS. B.

I have now drawn it asunder, and am going to blow a ball at one of the heated ends; but I must previously close it up, and flatten it with this little metallic instrument, otherwise the breath would pass through the tube without dilating any part of it. — Now, Caroline, will you blow strongly into the tube whilst the closed end is red hot.

EMILY.

You blowed too hard; for the ball suddenly dilated to a great size, and then burst in pieces.

MRS. B.

You will be more expert another time; but I must caution you, should you ever use this blow-pipe, to be very careful that the combustion of the alcohol does not go on with too great violence, for I have seen the flame sometimes dart out with such force as to reach the opposite wall of the room, and set the paint on fire. There is, however, no danger of the vessel bursting, as it is provided with a safety tube, which affords an additional vent for the vapour of alcohol when required.

The products of the combustion of alcohol consist in a great proportion of water, and a small quantity of carbonic acid. There is no smoke or fixed remains whatever. — How do you account for that, Emily?

EMILY.

I suppose that the oxygen which the alcohol absorbs in burning, converts its hydrogen into water and its carbon into carbonic acid gas, and thus it is completely consumed.

MRS. B.

Very well. — *Ether*, the lightest of all fluids, and with which you are well acquainted, is obtained from alcohol, of which it forms the lightest and most volatile part.

EMILY.

Ether, then, is to alcohol, what alcohol is to brandy?

MRS. B.

No: there is an essential difference. In order to obtain alcohol from brandy, you need only deprive the latter of its water; but for the formation of ether, the alcohol must be decomposed, and one of its constituents partly subtracted. I leave you to guess which of them it is —

EMILY.

It cannot be hydrogen, as ether is more volatile than alcohol, and hydrogen is the lightest of all its ingredients: nor do I suppose that it can be oxygen, as alcohol contains so small a proportion of that principle; it is, therefore, most probably, carbon, a diminution of which would not fail to render the new compound more volatile.

MRS. B.

You are perfectly right. The formation of ether consists simply in subtracting from the alcohol a certain proportion of carbon; this is effected by the action of the sulphuric, nitric, or muriatic acids, on alcohol. The acid and carbon remain at the bottom of the vessel, whilst the decarbonised alcohol flies off in the form of a condensable vapour, which is ether.

Ether is the most inflammable of all fluids, and burns at so slow a temperature that the heat evolved during its combustion is more than is required for its support, so that a quantity of ether is volatilised, which takes fire, and gradually increases the violence of the combustion.

Sir Humphry Davy has lately discovered a very singular fact respecting the vapour of ether. If a few drops of ether be poured into a wine-glass, and a fine platina wire, heated almost to redness, be held suspended in the glass, close to the surface of the ether, the wire soon becomes intensely red-hot, and remains so for any length of time. We may easily try the experiment. . . . ,

CAROLINE.

How very curious! The wire is almost white hot, and a pungent smell rises from the glass. Pray how is this accounted for?

MRS. B.

This is owing to a very peculiar property of the vapour of ether, and indeed of many other combustible gaseous bodies. At a certain temperature lower than that of ignition, these vapours undergo a slow and imperfect combustion, which does not give rise, in any sensible degree, to the phenomena of light and flame, and yet extricates a quantity of caloric sufficient to react upon the wire and make it red-hot, and the wire in its turn keeps up the effect as long as the emission of vapour continues.

CAROLINE.

But why should not an iron or silver wire produce the same effect?

MRS. B.

Because either iron or silver, being much better conductors of heat than platina, the heat is carried off too fast by those metals to allow the accumulation of caloric necessary to produce the effect in question.

Ether is so light that it evaporates at the common temperature of the atmosphere; it is therefore necessary to keep it confined by a well ground glass stopper. No degree of cold known has ever frozen it.

CAROLINE.

Is it not often taken medicinally?

MRS. B.

Yes; it is one of the most effectual antispasmodic medicines, and the quickness of its effects, as such, probably depends on its being instantly converted into vapour by the heat of the stomach, through the intervention of which it acts on the nervous system. But the frequent use of ether, like that of spirituous liquors, becomes prejudicial, and, if taken to excess, it produces effects similar to those of intoxication.

We may now take our leave of the vinous fermentation, of which, I hope, you have acquired a clear idea; as well as of the several products that are derived from it.

CAROLINE.

Though this process appears, at first sight, so much complicated, it may, I think, be summed up in a few words, as it consists in the conversion of sugar and fermentable bodies into alcohol and carbonic acid, which give rise both to the formation of wine, and of all kinds of spirituous liquors.

MRS. B.

We shall now proceed to the *acetous fermentation*, which is thus called, because it converts wine into vinegar, by the formation of the acetous acid, which is the basis or radical of vinegar.

CAROLINE.

But is not the acidifying principle of the acetous acid the same as that of all other acids, oxygen?

MRS. B.

Certainly; and on that account the contact of air is essential to this fermentation, as it affords the necessary supply of oxygen. Vinegar, in order to obtain pure acetous acid from it, must be distilled and rectified by certain processes.

EMILY.

But pray, Mrs. B., is not the acetous acid frequently formed without this fermentation taking place? Is it not, for instance, contained in acid fruits, and in every substance that becomes sour?

MRS. B.

No, not in fruits; you confound it with the citric, the malic, the oxalic, and other vegetable acids, to which living vegetables owe their acidity. But whenever a vegetable substance turns sour, after it has ceased to live, the acetous acid is developed by means of the acetous fermentation, in which the substance advances a step towards its final decomposition.

Amongst the various instances of acetous fermentation, that of bread is usually classed.

CAROLINE.

But the fermentation of bread is produced by yeast; how does that effect it?

MRS. B.

It is found by experience that any substance that has already undergone a fermentation, will readily excite it in one that is susceptible of that process. If, for instance, you mix a little vinegar with wine, that is intended to be acidified, it will absorb oxygen more rapidly, and the process be completed much sooner, than if left to ferment spontaneously. Thus yeast, which is a product of the fermentation of beer, is used to excite and accelerate the fermentation of malt, which is to be converted into beer, as well as that of paste which is to be made into bread.

CAROLINE.

But if bread undergoes the acetous fermentation, why is it not sour?

MRS. B.

It acquires a certain savour which corrects the heavy insipidity of flour, and may be reckoned a first degree of acidification; or if the process were carried further, the bread would become decidedly acid.

There are, however, some chemists who do not

consider the fermentation of bread as being of the acetous kind, but suppose that it is a process of fermentation peculiar to that substance.

The *putrid fermentation* is the final operation of Nature, and her last step towards reducing organised bodies to their simplest combinations. All vegetables spontaneously undergo this fermentation after death, provided there be a sufficient degree of heat and moisture, together with access of air; for it is well known that dead plants may be preserved by drying, or by the total exclusion of air.

CAROLINE.

But do dead plants undergo the other fermentation previous to this last; or do they immediately suffer the putrid fermentation?

MRS. B.

That depends on a variety of circumstances, such as the degrees of temperature and of moisture, the nature of the plant itself, &c. But if you were carefully to follow and examine the decomposition of plants from their death to their final dissolution, you would generally find a sweetness developed in the seeds, and a spirituous flavour in the fruits (which have undergone the saccharine fermentation), previous to the total disorganisation and separation of the parts.

EMILY.

I have sometimes remarked a kind of spirituous taste in fruits that were over ripe, especially oranges; and this was just before they became rotten.

MRS. B.

It was then the vinous fermentation which had succeeded the saccharine, and had you followed up these changes attentively, you would probably have found the spirituous taste followed by acidity, previous to the fruit passing to the state of putrefaction.

When the leaves fall from the trees in autumn, they do not (if there is no great moisture in the atmosphere) immediately undergo a decomposition, but are first dried and withered; as soon, however, as the rain sets in, fermentation commences, their gaseous products are imperceptibly evolved into the atmosphere, and their fixed remains mixed with their kindred earth.

Wood, when exposed to moisture, also undergoes the putrid fermentation and becomes rotten.

EMILY.

But I have heard that the *dry rot*, which is so liable to destroy the beams of houses, is prevented by a current of air; and yet you said that air was essential to the putrid fermentation?

MRS. B.

True; but it must not be in such a proportion to the moisture as to dissolve the latter, and this is generally the case when the rotting of wood is prevented or stopped by the free access of air. What is commonly called dry rot, however, is not I believe a true process of putrefaction. It is supposed to depend on a peculiar kind of vegetation, which, by feeding on the wood, gradually destroys it.

Straw and all other kinds of vegetable matter undergo the putrid fermentation more rapidly when mixed with animal matter. Much heat is evolved during this process, and a variety of volatile products are disengaged, as carbonic acid and hydrogen gas, the latter of which is frequently either sulphurated or phosphorated.— When all these gases have been evolved, the fixed products, consisting of carbon, salts, potash, &c. form a kind of vegetable earth, which makes very fine manure, as it is composed of those elements which form the immediate materials of plants.

CAROLINE.

Pray are not vegetables sometimes preserved from decomposition by petrification? I have seen very curious specimens of petrified vegetables, in which state they perfectly preserve their form and organisation, though in appearance they are changed to stone.

MRS. B.

That is a kind of metamorphosis, which, now that you are tolerably well versed in the history of mineral and vegetable substances, I leave to your judgment to explain. Do you imagine that vegetables can be converted into stone?

EMILY.

No, certainly; but they might perhaps be changed to a substance in appearance resembling stone.

MRS. B.

It is not so, however, with the substances that are called petrified vegetables; for these are really stone, and generally of the hardest kind, consisting chiefly of silex. The case is this: when a vegetable is buried under water, or in wet earth, it is slowly and gradually decomposed. As each successive particle of the vegetable is destroyed, its place is supplied by a particle of siliceous earth, conveyed thither by the water. In the course of time the vegetable is entirely destroyed, but the silex has completely replaced it, having assumed its form and apparent texture, as if the vegetable itself were changed to stone.

CAROLINE.

That is very curious! and I suppose that petrified animal substances are of the same nature?

MRS. B.

Precisely. It is equally impossible for either animal or vegetable substances to be converted into stone. They may be reduced, as we find they are, by decomposition, to their constituent elements, but cannot be changed to elements, which do not enter into their composition.

There are, however, circumstances which frequently prevent the regular and final decomposition of vegetables; as, for instance, when they are buried either in the sea, or in the earth, where they cannot undergo the putrid fermentation for want of air. In these cases they are subject to a peculiar change, by which they are converted into a new class of compounds, called *bitumens.*

CAROLINE.

These are substances I never heard of before.

MRS. B.

You will find, however, that some of them are very familiar to you. Bitumens are vegetables so far decomposed as to retain no organic appearance; but their origin is easily detected by their oily nature, their combustibility, the products of their analysis, and the impressions of the forms of leaves, grains, fibres of wood, and even of animals, which they frequently bear.

They are sometimes of an oily liquid consistence, as the substance called *naptha*, in which we preserved potassium; it is a fine transparent colourless fluid, that issues out of clays in some parts of Persia. But more frequently bitumens are solid, as *asphaltum*, a smooth, hard, brittle substance, which easily melts, and forms, in its liquid state, a beautiful dark brown colour for oil painting. *Jet*, which is of a still harder texture, is a peculiar bitumen, susceptible of so fine a polish, that it is used for many ornamental purposes.

Coal is also a bituminous substance, to the composition of which both the mineral and animal kingdoms seem to concur. This most useful mineral appears to consist chiefly of vegetable matter, mixed with the remains of marine animals and marine salts, and occasionally containing a quantity of sulphuret of iron, commonly called pyrites.

EMILY.

It is, I suppose, the earthly, the metallic, and the saline parts of coals, that compose the cinders or fixed products of their combustion; whilst the hydrogen and carbon, which they derive from vegetables, constitute their volatile products.

CAROLINE.

Pray is not *coke*, (which I have heard is much used in some manufactures,) also a bituminous substance?

MRS. B.

No; it is a kind of fuel artificially prepared from coals. It consists of coals reduced to a substance analogous to charcoal, by the evaporation of their bituminous parts. Coke, therefore, is composed of carbon, with some earthy and saline ingredients.

Succin, or *yellow amber*, is a bitumen which the ancients called *electrum*, from whence the word electricity is derived, as that substance is peculiarly, and was once supposed to be exclusively, electric. It is found either deeply buried in the bowels of the earth, or floating on the sea, and is supposed to be a resinous body which has been acted on by sulphuric acid, as its analysis shows it to consist of an oil and an acid. The oil is called *oil of amber*, the acid the *succinic*.

EMILY.

That oil I have sometimes used in painting, as it is reckoned to change less than the other kinds of oils.

MRS. B.

The last class of vegetable substances that have changed their nature are *fossil-wood*, *peat*, and *turf*. These are composed of wood and roots of shrubs, that are partly decomposed by being exposed to moisture under ground, and yet, in some measure, preserve their form and organic appearance. The peat, or black earth of the moors, retains but few vestiges of the roots to which it owes its richness and combustibility, these substances being in the course of time reduced to the state of vegetable earth. But in turf the roots of plants are still discernible, and it equally answers the purpose of fuel. It is the combustible used by the poor in heathy countries, which supply it abundantly.

It is too late this morning to enter upon the history of vegetation. We shall reserve this subject, therefore, for our next interview, when I expect that it will furnish us with ample matter for another conversation.

CONVERSATION XXII.

HISTORY OF VEGETATION.

MRS. B.

THE VEGETABLE KINGDOM may be considered as the link which unites the mineral and animal creation into one common chain of beings; for it is through the means of vegetation alone that mineral substances are introduced into the animal system, since, generally speaking, it is from vegetables that all animals ultimately derive their sustenance.

CAROLINE.

I do not understand that; the human species subsists as much on animal as on vegetable food, and there are some carnivorous animals that will eat only animal food.

MRS. B.

That is true; but you do not consider that those that live on animal food, derive their sustenance equally, though not so immediately, from vegetables. The meat that we eat is formed from the herbs of the field, and the prey of carnivorous animals proceeds, either directly or indirectly, from the same source. It is, therefore, through this channel that the simple elements become a part of the animal frame. We should in vain attempt to derive nourishment from carbon, hydrogen, and oxygen, either in their separate state, or combined in the mineral kingdom; for it is only by being united in the form of vegetable combination, that they become capable of conveying nourishment.

EMILY.

Vegetation, then, seems to be the method which Nature employs to prepare the food of animals?

MRS. B.

That is certainly its principal object. The vegetable creation does not exhibit more wisdom in that admirable system of organisation, by which it is enabled to answer its own immediate ends of preservation, nutrition, and propagation, than in its grand and ultimate object of forming those arrangements and combinations of principles, which are so well adapted for the nourishment of animals.

EMILY.

But I am very curious to know whence vegetables obtain those principles which form their immediate materials?

MRS. B.

This is a point on which we are yet so much in the dark, that I cannot hope fully to satisfy your curiosity; but what little I know on this subject, I will endeavour to explain to you.

The soil, which, at first view, appears to be the aliment of vegetables, is found, on a closer investigation, to be little more than the channel through which they receive their nourishment; so that it is very possible to rear plants without any earth or soil.

CAROLINE.

Of that we have an instance in the hyacinth and other bulbous roots, which will grow and blossom beautifully in glasses of water. But I confess I should think it would be difficult to rear trees in a similar manner.

MRS. B.

No doubt it would, as it is the burying of the roots in the earth that supports the stem of the tree. But this office, besides that of affording a vehicle for food, is far the most important part which the earthy portion of the soil performs in the process of vegetation; for we can discover, by analysis, but an extremely small proportion of earth in vegetable compounds.

CAROLINE.

But if earths do not afford nourishment, why is it necessary to be so attentive to the preparation of the soil?

MRS. B.

In order to impart to it those qualities which render it a proper vehicle for the food of the plant. Water is the chief nourishment of vegetables; if, therefore, the soil be too sandy, it will not retain a quantity of water sufficient to supply the roots of the plants. If, on the contrary, it abound too much with clay, the water will lodge in such quantities as to threaten a decomposition of the roots. Calcareous soils are, upon the whole, the most favourable to the growth of plants: soils are, therefore, usually improved by chalk, which, you may recollect, is a carbonat of lime. Different vegetables, however, require different kinds of soils. Thus rice demands a moist retentive soil; potatoes a soft sandy soil; wheat a firm and rich soil. Forest trees grow better in fine sand than in a stiff clay; and a light ferruginous soil is best suited to fruit-trees.

CAROLINE.

But pray what is the use of manuring the soil?

MRS. B.

Manure consists of all kinds of substances, whether of vegetable or animal origin, which have undergone the putrid fermentation, and are consequently decomposed, or nearly so, into their elementary principles. And it is requisite that these vegetable matters should be in a state of decay, or approaching decomposition. The addition of calcareous earth, in the state of chalk or lime, is beneficial to such soils, as it accelerates the dissolution of vegetable bodies. Now, I ask you, what is the utility of supplying the soil with these decomposed substances?

CAROLINE.

It is, I suppose, in order to furnish vegetables with the principles which enter into their composition. For manures not only contain carbon, hydrogen, and oxygen, but by their decomposition supply the soil with these principles in their elementary form.

MRS. B.

Undoubtedly; and it is for this reason that the finest crops are produced in fields that were formerly covered with woods, because their soil is composed of a rich mould, a kind of vegetable earth, which abounds in those principles.

EMILY.

This accounts for the plentifulness of the crops

produced in America, where the country was but a few years since covered with wood.

CAROLINE.

But how is it that animal substances are reckoned to produce the best manure? Does it not appear much more natural that the decomposed elements of vegetables should be the most appropriate to the formation of new vegetables?

MRS. B.

The addition of a much greater proportion of nitrogen, which constitutes the chief difference between animal and vegetable matter, renders the composition of the former more complicated, and consequently more favourable to decomposition. The use of animal substances is chiefly to give the first impulse to the fermentation of the vegetable ingredients that enter into the composition of manures. The manure of a farm-yard is of that description; but there is scarcely any substance susceptible of undergoing the putrid fermentation that will not make good manure. The heat produced by the fermentation of manure is another circumstance which is extremely favourable to vegetation; yet this heat would be too great if the manure was laid on the ground during the height of fermentation; it is used in this state only for hot-beds, to produce melons, cucumbers, and such vegetables as require a very high temperature.

CAROLINE.

A difficulty has just occurred to me which I do not know how to remove. Since all organised bodies are, in the common course of nature, ultimately reduced to their elementary state, they must necessarily in that state enrich the soil, and afford food for vegetation. How is it, then, that agriculture, which cannot increase the quantity of those elements that are required to manure the earth, can increase its produce so wonderfully as is found to be the case in all cultivated countries?

MRS. B.

It is by suffering none of these decaying bodies to be dissipated, but in applying them duly to the soil. It is by a judicious preparation of the soil, which consists in fitting it either for the general purposes of vegetation, or for that of the particular seed which is to be sown. Thus, if the soil be too wet, it may be drained; if too loose and sandy, it may be rendered more consistent and retentive of water by the addition of clay or loam; it may be enriched by chalk, or any kind of calcareous earth. On soils thus improved, manures will act with double efficacy, and if attention be paid to spread them on the ground at a proper season of the year, to mix them with the soil so that they may be generally diffused through it, to destroy the weeds which might appropriate these nutritive principles to their own use, to remove the stones which would impede the growth of the plant, &c. we may obtain a produce an hundred fold more abundant than the earth would spontaneously supply.

EMILY.

We have a very striking instance of this in the scanty produce of uncultivated commons, compared to the rich crops of meadows which are occasionally manured.

CAROLINE.

But, Mrs. B., though experience daily proves the advantage of cultivation, there is still a difficulty which I cannot get over. A certain quantity of elementary principles exist in nature, which it is not in the power of man either to augment or diminish. Of these principles you have taught us that both the animal and vegetable creation are composed. Now the more of them is taken up by the vegetable kingdom, the less, it would seem, will remain for animals; and, therefore, the more populous the earth becomes, the less it will produce.

MRS. B.

Your reasoning is very plausible; but expe-

PLATE XV.

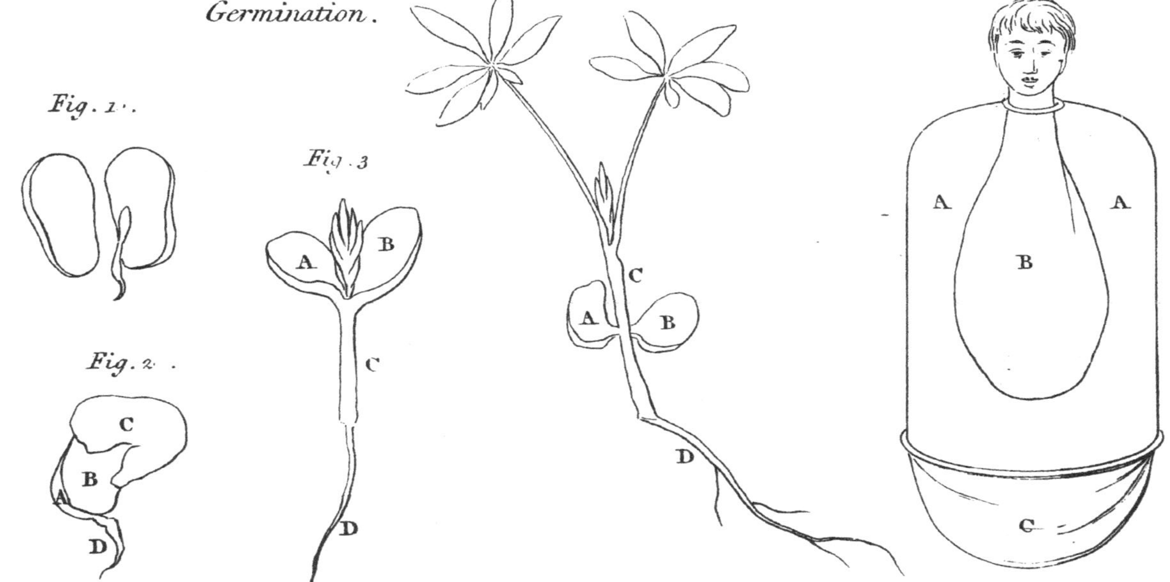

Fig 1 & 2 A.B Cotyledons.—C Envelope.—D Radicle.—Fig. 3. A.B Cotyledons.—C Plumula.—D Radicle.—Fig. 4. A.B. Cotyledons. C Plumula.—D Radicle.—Fig. 5. A.A Glass Bell.—B Bladder representing the lungs. C Bladder representing the Diaphragm.

Drawn by the Author. *Published by Longman & Co. Oct.r 2nd 1809.* *Engraved by Lowry*

rience every where contradicts the inference you would draw from it: for we find that the animal and vegetable kingdoms, instead of thriving, as you would suppose, at each other's expense, always increase and multiply together. For you should recollect that animals can derive the elements of which they are formed only through the medium of vegetables. And you must allow that your conclusion would be valid only if every particle of the several principles that could possibly be spared from other purposes were employed in the animal and vegetable creations. Now we have reason to believe that a much greater proportion of these principles than is required for such purposes remains either in an elementary state, or engaged in a less useful mode of combination in the mineral kingdom. Possessed of such immense resources as the atmosphere and the waters afford us, for oxygen, hydrogen, and carbon, so far from being in danger of working up all our simple materials, we cannot suppose that we shall ever bring agriculture to such a degree of perfection as to require the whole of what these resources could supply.

Nature, however, in thus furnishing us with an inexhaustible stock of raw materials, leaves it in some measure to the ingenuity of man to appropriate them to its own purposes. But, like a kind parent, she stimulates him to exertion, by setting the example and pointing out the way. For it is on the operations of nature that all the improvements of art are founded. The art of agriculture consists, therefore, in discovering the readiest method of obtaining the several principles, either from their grand sources, air and water, or from the decomposition of organised bodies; and in appropriating them in the best manner to the purposes of vegetation.

EMILY.

But, among the sources of nutritive principles, I am surprised that you do not mention the earth itself, as it contains abundance of coals, which are chiefly composed of carbon.

MRS. B.

Though coals abound in carbon, they cannot, on account of their hardness and impermeable texture, be immediately subservient to the purposes of vegetation.

EMILY.

No; but by their combustion carbonic acid is produced; and this entering into various combinations on the surface of the earth, may, perhaps, assist in promoting vegetation.

MRS. B.

Probably it may in some degree; but at any

rate the quantity of nourishment which vegetables may derive from that source can be but very trifling, and must entirely depend on local circumstances.

CAROLINE.

Perhaps the smoky atmosphere of London is the cause of vegetation being so forward and so rich in its vicinity?

MRS. B.

I rather believe that this circumstance proceeds from the very ample supply of manure, assisted, perhaps, by the warmth and shelter which the town affords. Far from attributing any good to the smoky atmosphere of London, I confess I like to anticipate the time when we shall have made such progress in the art of managing combustion, that every particle of carbon will be consumed, and the smoke destroyed at the moment of its production. We may then expect to have the satisfaction of seeing the atmosphere of London as clear as that of the country. — But to return to our subject: I hope that you are now convinced that we shall not easily experience a deficiency of nutritive elements to fertilise the earth, and that, provided we are but industrious in applying them to the best advantage by improving the art of agriculture, no limits can be assigned to the fruits that we may expect to reap from our labours.

CAROLINE.

Yes; I am perfectly satisfied in that respect, and I can assure you that I feel already much more interested in the progress and improvement of agriculture.

EMILY.

I have frequently thought that the culture of the land was not considered as a concern of sufficient importance. Manufactures always take the lead; and health and innocence are frequently sacrificed to the prospect of a more profitable employment. It has often grieved me to see the poor manufacturers crowded together in close rooms, and confined for the whole day to the most uniform and sedentary employment, instead of being engaged in that innocent and salutary kind of labour, which Nature seems to have assigned to man for the immediate acquirement of comfort, and for the preservation of his existence. I am sure that you agree with me in thinking so, Mrs. B.?

MRS. B.

I am entirely of your opinion, my dear, in regard to the importance of agriculture; but as the conveniences of life, which we are all enjoying, are not derived merely from the soil, I am far from wishing to depreciate manufactures. Besides, as the labour of one man is sufficient to produce food for several, those whose industry is not required in tillage must do something in return for the food that is provided for them. They exchange, consequently, the accommodations for the necessaries of life. Thus the carpenter and the weaver lodge and clothe the peasant, who supplies them with their daily bread. The greater stock of provisions, therefore, which the husbandman produces, the greater is the quantity of accommodation which the artificer prepares. Such are the happy effects which naturally result from civilised society. It would be wiser, therefore, to endeavour to improve the situation of those who are engaged in manufactures, than to indulge in vain declamations on the hardships to which they are too frequently exposed.

But we must not yet take our leave of the subject of agriculture; we have prepared the soil, it remains for us now to sow the seed. In this operation we must be careful not to bury it too deep in the ground, as the access of air is absolutely necessary to its germination; the earth must, therefore, lie loose and light over it, in order that the air may penetrate. Hence the use of ploughing and digging, harrowing and raking, &c. A certain degree of heat and moisture, such as usually takes place in the spring, is likewise necessary.

CAROLINE.

One would imagine you were going to describe

the decomposition of an old plant, rather than the formation of a new one; for you have enumerated all the requisites of fermentation.

MRS. B.

Do you forget, my dear, that the young plant derives its existence from the destruction of the seed, and that it is actually by the saccharine fermentation that the latter is decomposed?

CAROLINE.

True; I wonder that I did not recollect that. The temperature and moisture required for the germination of the seed is then employed in producing the saccharine fermentation within it?

MRS. B.

Certainly. But, in order to understand the nature of germination, you should be acquainted with the different parts of which the seed is composed. The external covering or envelope contains, besides the germ of the future plant, the substance which is to constitute its first nourishment; this substance, which is called the *parenchyma*, consists of fecula, mucilage, and oil, as we formerly observed.

The seed is generally divided into two compartments, called *lobes*, or *cotyledons*, as is exemplified by this bean (PLATE XV. Fig. 1.) — the dark-coloured kind of string which divides the lobes is called the *radicle*, as it forms the root of the plant, and it is from a contiguous substance, called *plumula*, which is enclosed within the lobes, that the stem arises. The figure and size of the seed depend very much upon the cotyledons; these vary in number in different seeds; some have only one, as wheat, oats, barley, and all the grasses; some have three, others six. But most seeds, as, for instance, all the varieties of beans, have two cotyledons. When the seed is buried in the earth, at any temperature above 40 degrees, it imbibes water, which softens and swells the lobes; it then absorbs oxygen, which combines with some of its carbon, and is returned in the form of carbonic acid. This loss of carbon increases the comparative proportion of hydrogen and oxygen in the seed, and excites the saccharine fermentation, by which the parenchymatous matter is converted into a kind of sweet emulsion. In this form it is carried into the radicle by vessels appropriated to that purpose; and in the mean time, the fermentation having caused the seed to burst, the cotyledons are rent asunder, the radicle strikes into the ground and becomes the root of the plant, and hence the fermented liquid is conveyed to the plumula, whose vessels have been previously distended by the heat of the fermentation. The plumula being thus swelled, as it were, by the emulsive fluid, raises itself and springs up to the surface of the earth, bearing with it the cotyledons, which, as soon as they come in contact with the air, spread themselves, and are transformed into leaves. — If we go into the garden, we shall probably find some seeds in the state which I have described —

EMILY.

Here are some lupines that are just making their appearance above ground.

MRS. B.

We shall take up several of them to observe their different degrees of progress in vegetation. Here is one that has but recently burst its envelope — do you see the little radicle striking downwards? (PLATE XV. Fig. 2.) In this the plumula is not yet visible. But here is another in a greater state of forwardness — the plumula, or stem, has risen out of the ground, and the cotyledons are converted into seed leaves. (PLATE XV. Fig. 3.)

CAROLINE.

These leaves are very thick and clumsy, and unlike the other leaves, which I perceive are just beginning to appear.

MRS. B.

It is because they retain the remains of the parenchyma, with which they still continue to nourish the young plant, as it has not yet sufficient roots and strength to provide for its sustenance from the soil. — But, in this third lupine (PLATE XIV. Fig. 4.), the radicle had sunk deep into the earth, and sent out several shoots, each of which is furnished with a mouth to suck up nourishment from the soil; the function of the original leaves, therefore, being no longer required, they are gradually decaying, and the plumula is become a regular stem, shooting out small branches, and spreading its foliage.

EMILY.

There seems to be a very striking analogy between a seed and an egg; both require an elevation of temperature to be brought to life; both at first supply with aliment the organised being which they produce; and as soon as this has attained sufficient strength to procure its own nourishment, the egg-shell breaks, whilst in the plant the seed-leaves fall off.

MRS. B.

There is certainly some resemblance between these processes; and when you become acquainted with animal chemistry, you will frequently be struck with its analogy to that of the vegetable kingdom.

As soon as the young plant feeds from the soil, it requires the assistance of leaves, which are the organs by which it throws off its super-abundant fluid; this secretion is much more plentiful in the vegetable than in the animal creation, and the great extent of surface of the foliage of plants is admirably calculated for carrying it on in sufficient quantities. This transpired fluid consists of little more than water. The sap, by this process, is converted into a liquid of greater consistence, which is fit to be assimilated to its several parts.

EMILY.

Vegetation, then, must be essentially injured by destroying the leaves of the plant?

MRS. B.

Undoubtedly; it not only diminishes the transpiration, but also the absorption by the roots; for the quantity of sap absorbed is always in proportion to the quantity of fluid thrown off by transpiration. You see, therefore, the necessity that a young plant should unfold its leaves as soon as it begins to derive its nourishment from the soil; and, accordingly, you will find that those lupines which have dropped their seed-leaves, and are no longer fed by the parenchyma, have spread their foliage, in order to perform the office just described.

But I should inform you that this function of transpiration seems to be confined to the upper surface of the leaves, whilst, on the contrary, the lower surface, which is more rough and uneven, and furnished with a kind of hair or down, is destined to absorb moisture, or such other ingredients as the plant derives from the atmosphere.

As soon as a young plant makes its appearance above ground, light, as well as air, becomes necessary to its preservation. Light is essential to the development of the colours, and to the thriving of the plant. You may have often observed what a predilection vegetables have for the light. If you make any plants grow in a room, they all spread their leaves, and extend their branches towards the windows.

CAROLINE.

And many plants close up their flowers as soon as it is dark.

EMILY.

But may not this be owing to the cold and dampness of the evening air?

MRS. B.

That does not appear to be the case; for in a course of curious experiments, made by Mr. Senebier, of Geneva, on plants which he reared by lamp-light, he found that the flowers closed their petals whenever the lamps were extinguished.

EMILY.

But pray, why is air essential to vegetation, plants do not breathe it like animals?

MRS. B.

At least not in the same manner; but they certainly derive some principles from the atmosphere, and yield others to it. Indeed, it is chiefly owing to the action of the atmosphere and the vegetable kingdom on each other, that the air continues always fit for respiration. But you will understand this better when I have explained the effect of water on plants.

I have said that water forms the chief nourishment of plants; it is the basis not only of the sap, but of all the vegetable juices. Water is the vehicle which carries into the plant the various salts and other ingredients required for the formation and support of the vegetable system. Nor is this all; part of the water itself is decomposed by the organs of the plant; the hydrogen becomes a constituent part of oil, of extract, of colouring matter, &c. whilst a portion of the oxygen enters into the formation of mucilage, of fecula, of sugar, and of vegetable acids. But the greater part of the oxygen, proceeding from the decomposition of the water, is converted into a gaseous state by the caloric disengaged from the hydrogen during its condensation in the formation of the vegetable materials. In this state the oxygen is transpired by the leaves of plants when exposed to the sun's rays. Thus you find that the decomposition of water, by the organs of the plant, is not only a means of supplying it with its chief ingredient, hydrogen, but at the same time of replenishing the atmosphere with oxygen, a principle which requires continual renovation, to make up for the great consumption of it occasioned by the numerous oxygenations, combustions, and respirations, that are constantly taking place on the surface of the globe.

EMILY.

What a striking instance of the harmony of nature.

MRS. B.

And how admirable the design of Providence, who makes every different part of the creation thus contribute to the support and renovation of each other!

But the intercourse of the vegetable and animal kingdoms through the medium of the atmosphere extends still further. Animals, in breathing, not only consume the oxygen of the air, but load it with carbonic acid, which, if accumulated in the atmosphere, would, in a short time, render it totally unfit for respiration. Here the vegetable kingdom again interferes; it attracts and decomposes the carbonic acid, retains the carbon for its own purposes, and returns the oxygen for ours.

CAROLINE.

How interesting this is! I do not know a more beautiful illustration of the wisdom which is displayed in the laws of nature.

MRS. B.

Faint and imperfect as are the ideas which our limited perceptions enable us to form of divine wisdom, still they cannot fail to inspire us with awe and admiration. What, then, would be our feelings, were the complete system of nature at once displayed before us! So magnificent a scene would probably be too great for our limited and imperfect comprehension, and it is no doubt among the wise dispensations of Providence, to veil the splendour of a glory with which we should be overpowered. But it is well suited to the nature of a rational being to explore, step by step, the works of the creation, to endeavour to connect them into harmonious systems; and, in a word, to trace in the chain of beings, the kindred ties and benevolent design which unites its various links, and secure its preservation.

CAROLINE.

But of what nature are the organs of plants which are endued with such wonderful powers?

MRS. B.

They are so minute that their structure, as well as the mode in which they perform their functions, generally elude our examination; but we may consider them as so many vessels or apparatus appropriated to perform, with the assistance of the principle of life, certain chemical processes, by means of which these vegetable compounds are generated. We may, however, trace the tannin, resins, gum, mucilage, and some other vegetable materials, in the organised arrangement of plants, in which they form the bark, the wood, the leaves, flowers, and seeds.

The *bark* is composed of the *epidermis*, the *parenchyma*, and the *cortical layers*.

The epidermis is the external covering of the plant. It is a thin transparent membrane, consisting of a number of slender fibres, crossing each other, and forming a kind of net-work. When of a white glossy nature, as in several species of trees, in the stems of corn and of seeds, it is composed of a thin coating of siliceous earth, which accounts for the strength and hardness of those long and slender stems. Sir H. Davy was led to the discovery of the siliceous nature of the epidermis of such plants, by observing the singular phenomenon of sparks of fire emitted by the collision of ratan canes with which two boys were fighting in a dark room. On analysing the epidermis of the cane, he found it to be almost entirely siliceous.

CAROLINE.

With iron then, a cane, I suppose, will strike fire very easily?

MRS. B.

I understand that it will. — In ever-greens the epidermis is mostly resinous, and in some few plants is formed of wax. The resin, from its want of affinity for water, tends to preserve the plant from the destructive effects of violent rains, severe climates, or inclement seasons, to which this species of vegetables is peculiarly exposed.

EMILY.

Resin must preserve wood just like a varnish, as it is the essential ingredient of varnishes?

MRS. B.

Yes; and by this means it prevents likewise all unnecessary expenditure of moisture.

The parenchyma is immediately beneath the epidermis; it is that green rind which appears when you strip a branch of any tree or shrub of its external coat of bark. The parenchyma is not confined to the stem or branches, but extends over every part of the plant. It forms the green matter of the leaves, and is composed of tubes filled with a peculiar juice.

The cortical layers are immediately in contact with the wood; they abound with tannin and gallic acid, and consist of small vessels through which the sap descends after being elaborated in the leaves. The cortical layers are annually renewed, the old bark being converted into wood.

EMILY.

But through what vessels does the sap ascend?

MRS. B.

That function is performed by the tubes of the alburnum, or wood, which is immediately beneath the cortical layers. The wood is composed of woody fibre, mucilage, and resin. The fibres are disposed in two ways; some of them longitudinally, and these form what is called the silver grain of the wood. The others, which are concentric, are called the spurious grain. These last are disposed in layers, from the number of which the age of the tree may be computed, a new one being produced annually by the conversion of the bark into wood. The oldest, and consequently most internal part of the alburnum, is called heart-wood; it appears to be dead, at least no vital functions are discernible in it. It is through the tubes of the living alburnum that the sap rises. These, therefore, spread into the leaves, and there communicate with the extremities of the vessels of the cortical layers, into which they pour their contents.

CAROLINE.

Of what use, then, are the tubes of the parenchyma, since neither the ascending nor descending sap passes through them?

MRS. B.

They are supposed to perform the important function of secreting from the sap the peculiar juices from which the plant more immediately derives its nourishment. These juices are very conspicuous, as the vessels which contain them are much larger than those through which the sap circulates. The peculiar juices of plants differ much in their nature, not only in different species of vegetables, but frequently in different parts of the same individual plant: they are sometimes saccharine, as in the sugar-cane, sometimes resinous, as in firs and evergreens, sometimes of a milky appearance, as in the laurel.

EMILY.

I have often observed, that in breaking a young

shoot, or in bruising a leaf of laurel, a milky juice will ooze out in great abundance.

MRS. B.

And it is by making incisions in the bark that pitch, tar, and turpentine are obtained from fir-trees. The durability of this species of wood is chiefly owing to the resinous nature of its peculiar juices. The volatile oils have, in a great measure, the same preservative effects, as they defend the parts, with which they are connected, from the attack of insects. This tribe seems to have as great an aversion to perfumes, as the human species have delight in them. They scarcely ever attack any odoriferous parts of plants, and it is not uncommon to see every leaf of a tree destroyed by a blight, whilst the blossoms remain untouched. Cedar, sandal, and all aromatic woods, are on this account of great durability.

EMILY.

But the wood of the oak, which is so much esteemed for its durability, has, I believe, no smell. Does it derive this quality from its hardness alone?

MRS. B.

Not entirely; for the chesnut, though considerably harder and firmer than the oak, is not so lasting. The durability of the oak is, I believe, in a great measure owing to its having very little heart-wood, the alburnum preserving its vital functions longer than in other trees.

CAROLINE.

If incisions are made into the alburnum and cortical layers, may not the ascending and descending sap be procured in the same manner as the peculiar juice is from the vessels of the parenchyma?

MRS. B.

Yes; but in order to obtain specimens of these fluids, in any quantity, the experiment must be made in the spring, when the sap circulates with the greatest energy. For this purpose a small bent glass tube should be introduced into the incision, through which the sap may flow without mixing with any of the other juices of the tree. From the bark the sap will flow much more plentifully than from the wood, as the ascending sap is much more liquid, more abundant, and more rapid in its motion than that which descends; for the latter having been deprived by the operation of the leaves of a considerable part of its moisture, contains a much greater proportion of solid matter, which retards its motion. It does not appear that there is any excess of descending sap, as none ever exudes from the roots of plants; this process, therefore, seems to be carried on only in proportion to the wants of the plant, and the sap descends no further, and in no greater quantity, than is required to nourish the several organs. Therefore, though the sap rises and descends in the plant, it does not appear to undergo a real circulation.

The last of the organs of plants is the *flower*, or *blossom*, which produces the *fruits* and *seed*. These may be considered as the ultimate purpose of nature in the vegetable creation. From fruits and seeds animals derive both a plentiful source of immediate nourishment, and an ample provision for the reproduction of the same means of subsistence.

The seed which forms the final product of mature plants, we have already examined as constituting the first rudiments of future vegetation.

These are the principal organs of vegetation, by means of which the several chemical processes which are carried on during the life of the plant are performed.

EMILY.

But how are the several principles which enter into the composition of vegetables so combined by the organs of the plant as to be converted into vegetable matter?

MRS. B.

By chemical processes, no doubt; but the apparatus in which they are performed is so extremely minute as completely to elude our examination. We can form an opinion, therefore, only by the result of these operations. The sap is evidently composed of water, absorbed by the roots, and holding in solution the various principles which it derives from the soil. From the roots the sap ascends through the tubes of the alburnum into the stem, and thence branches out to every extremity of the plant. Together with the sap circulates a certain quantity of carbonic acid, which is gradually disengaged from the former by the internal heat of the plant.

CAROLINE.

What! have vegetables a peculiar heat, analogous to animal heat?

MRS. B.

It is a circumstance that has long been suspected; but late experiments have decided beyond a doubt that vegetable heat is considerably above that of unorganised matter in winter, and below it in summer. The wood of a tree is about sixty degrees, when the thermometer is seventy or eighty degrees. And the bark, though so much exposed, is seldom below forty in winter.

It is from the sap, after it has been elaborated by the leaves, that vegetables derive their nourishment; in its progress through the plant from the leaves to the roots, it deposits in the several sets of vessels with which it communicates, the materials on which the growth and nourishment of each plant depends. It is thus that the various peculiar juices, saccharine, oily, mucous, acid, and colouring, are formed; as also the more solid parts, fecula, woody fibre, tannin, resins, concrete salts: in a word, all the immediate materials of vegetables, as well as the organised parts of plants, which latter, besides the power of secreting these from the sap for the general purpose of the plant, have also that of applying them to their own particular nourishment.

EMILY.

But why should the process of vegetation take place only at one season of the year, whilst a total inaction prevails during the other?

MRS. B.

Heat is such an important chemical agent, that its effect, as such, might perhaps alone account for the impulse which the spring gives to vegetation. But, in order to explain the mechanism of that operation, it has been supposed that the warmth of the spring dilates the vessels of plants, and produces a kind of vacuum, into which the sap (which had remained in a state of inaction in the trunk during the winter) rises: this is followed by the ascent of the sap contained in the roots, and room is thus made for fresh sap, which the roots, in their turn, pump up from the soil. This process goes on till the plant blossoms and bears fruit, which terminates its summer career: but when the cold weather sets in, the fibres and vessels contract, the leaves wither, and are no longer able to perform their office of transpiration; and, as this secretion stops, the roots cease to absorb sap from the soil. If the plant be an annual, its life then terminates; if not, it remains in a state of torpid inaction during the winter; or the only internal motion that takes place is that of a small quantity of resinous juice, which slowly rises from the stem into the branches, and enlarges their buds during the winter.

CAROLINE.

Yet, in evergreens, vegetation must continue throughout the year.

MRS. B.

Yes; but in winter it goes on in a very imperfect manner, compared to the vegetation of spring and summer.

We have dwelt much longer on the history of vegetable chemistry than I had intended; but we have at length, I think, brought the subject to a conclusion.

CAROLINE.

I rather wonder that you did not reserve the account of the fermentations for the conclusion; for the decomposition of vegetables naturally follows their death, and can hardly, it seems, be introduced with so much propriety at any other period.

MRS. B.

It is difficult to determine at what point precisely it may be most eligible to enter on the history of vegetation; every part of the subject is so closely connected, and forms such an uninterrupted chain, that it is by no means easy to divide it. Had I begun with the germination of the seed, which, at first view, seems to be the most proper arrangement, I could not have explained the nature and fermentation of the seed, or have described the changes which manure must undergo, in order to yield the vegetable elements. To understand the nature of germination, it is necessary, I think, previously to decompose the parent plant, in order to become acquainted with the materials required for that purpose. I hope, therefore, that, upon second consideration, you will find that the order which I have adopted, though apparently less correct, is in fact the best calculated for the elucidation of the subject.

CONVERSATION XXIII.

ON THE COMPOSITION OF ANIMALS.

MRS. B.

We are now come to the last branch of chemistry, which comprehends the most complicated order of compound beings. This is the animal creation, the history of which cannot but excite the highest degree of curiosity and interest, though we often fail in attempting to explain the laws by which it is governed.

EMILY.

But since all animals ultimately derive their nourishment from vegetables, the chemistry of this order of beings must consist merely in the conversion of vegetable into animal matter.

MRS. B.

Very true; but the manner in which this is effected is, in a great measure, concealed from our observation. This process is called *animalisation*, and is performed by peculiar organs. The difference of the animal and vegetable kingdoms does not however depend merely on a different arrangement of combinations. A new principle abounds in the animal kingdom, which is but rarely and in very small quantities found in vegetables; this is nitrogen. There is likewise in animal substances a greater and more constant proportion of phosphoric acid, and other saline matters. But these are not essential to the formation of animal matter.

CAROLINE.

Animal compounds contain, then, four fundamental principles; oxygen, hydrogen, carbon, and nitrogen?

MRS. B.

Yes; and these form the immediate materials of animals, which are *gelatine*, *albumen*, and *fibrine*.

EMILY.

Are those all? I am surprised that animals should be composed of fewer kinds of materials than vegetables; for they appear much more complicated in their organisation.

MRS. B.

Their organisation is certainly more perfect and intricate, and the ingredients that occasionally enter into their composition are more numerous. But notwithstanding the wonderful variety observable in the texture of the animal organs, we find that the original compounds, from which all the varieties of animal matter are derived, may be reduced to the three heads just mentioned. Animal substances being the most complicated of all natural compounds, are most easily susceptible of decomposition, as the scale of attractions increases in proportion to the number of constituent principles. Their analysis is, however, both difficult and imperfect; for as they cannot be examined in their living state, and are liable to alteration immediately after death, it is probable that, when submitted to the investigation of a chemist, they are always more or less altered in their combinations and properties, from what they were, whilst they made part of the living animal.

EMILY.

The mere diminution of temperature, which they experience by the privation of animal heat, must, I should suppose, be sufficient to derange the order of attractions that existed during life.

MRS. B.

That is one of the causes, no doubt: but there are many other circumstances which prevent us from studying the nature of living animal substances. We must therefore, in a considerable degree, confine our researches to the phenomena of these compounds in their inanimate state.

These three kinds of animal matter, gelatine, albumen, and fibrine, form the basis of all the various parts of the animal system; either solid, as the *skin*, *flesh*, *nerves*, *membranes*, *cartilages*, and *bones*; or fluid, as *blood*, *chyle*, *milk*, *mucus*, the *gastric* and *pancreatic juices*, *bile*, *perspiration*, *saliva*, *tears*, &c.

CAROLINE.

Is it not surprising that so great a variety of substances, and so different in their nature, should yet all arise from so few materials, and from the same original elements?

MRS. B.

The difference in the nature of various bodies depends, as I have often observed to you, rather on their state of combination, than on the materials of which they are composed. Thus, in considering the chemical nature of the creation in a general point of view, we observe that it is throughout composed of a very small number of elements. But when we divide it into the three kingdoms, we find that, in the mineral, the combinations seem to result from the union of elements casually brought together; whilst in the vegetable and animal kingdoms, the attractions are peculiarly and regularly produced by appropriate organs, whose action depends on the vital principle. And we may further observe, that by means of certain spontaneous changes and decompositions, the elements of one kind of matter become subservient to the reproduction of another; so that the three kingdoms are intimately connected, and constantly contributing to the preservation of each other.

EMILY.

There is, however, one very considerable class of elements, which seems to be confined to the mineral kingdom: I mean metals.

MRS. B.

Not entirely; they are found, though in very minute quantities, both in the vegetable and animal kingdoms. A small portion of earths and sulphur enters also into the composition of organised bodies. Phosphorus, however, is almost entirely confined to the animal kingdom; and nitrogen, but with few exceptions, is extremely scarce in vegetables.

Let us now proceed to examine the nature of the three principal materials of the animal system.

Gelatine, or *jelly*, is the chief ingredient of skin,

and of all the membranous parts of animals. It may be obtained from these substances, by means of boiling water, under the forms of glue, size, isinglass, and transparent jelly.

CAROLINE.

But these are of a very different nature; they cannot therefore be all pure gelatine.

MRS. B.

Not entirely, but very nearly so. Glue is extracted from the skin of animals. Size is obtained either from skin in its natural state, or from leather. Isinglass is gelatine procured from a particular species of fish; it is, you know, of this substance that the finest jelly is made, and this is done by merely dissolving the isinglass in boiling water, and allowing the solution to congeal.

EMILY.

The wine, lemon, and spices, are, I suppose, added only to flavour the jelly?

MRS. B.

Exactly so.

CAROLINE.

But jelly is often made of hartshorn shavings, and of calves' feet; do these substances contain gelatine?

MRS. B.

Yes. Gelatine may be obtained from almost any animal substance, as it enters more or less into the composition of all of them. The process for obtaining it is extremely simple, as it consists merely in boiling the substance that contains it with water. The gelatine dissolves in water, and may be attained of any degree of consistence or strength, by evaporating this solution. Bones in particular produce it very plentifully, as they consist of phosphat of lime combined or cemented by gelatine. Horns, which are a species of bone, will yield abundance of gelatine. The horns of the hart are reckoned to produce gelatine of the finest quality; they are reduced to the state of shavings in order that the jelly may be more easily extracted by the water. It is of hartshorn shavings that the jellies for invalids are usually made, as they are of very easy digestion.

CAROLINE.

It appears singular that hartshorn, which yields such a powerful ingredient as ammonia, should at the same time produce so mild and insipid a substance as jelly?

MRS. B.

And (what is more surprising) it is from the gelatine of bones that ammonia is produced. You must observe, however, that the processes by which these two substances are obtained from bones are very different. By the simple action of water and heat, the gelatine is separated; but in order to procure the ammonia, or what is commonly called hartshorn, the bones must be distilled, by which means the gelatine is decomposed, and hydrogen and nitrogen combined in the form of ammonia. So that the first operation is a mere separation of ingredients, whilst the second requires a chemical decomposition.

CAROLINE.

But when jelly is made from hartshorn shavings, what becomes of the phosphat of lime which constitutes the other part of bones?

MRS. B.

It is easily separated by straining. But the jelly is afterwards more perfectly purified, and rendered transparent, by adding white of egg, which being coagulated by heat, rises to the surface along with any impurities.

EMILY.

I wonder that bones are not used by the common people to make jelly; a great deal of wholesome nourishment, might, I should suppose, be procured from them, though the jelly would perhaps not be quite so good as if made from hartshorn shavings?

MRS. B.

There is a prejudice among the poor against a species of food that is usually thrown to the dogs; and as we cannot expect them to enter into chemical considerations, it is in some degree excusable. Besides, it requires a prodigious quantity of fuel to dissolve bones and obtain the gelatine from them.

The solution of bones in water is greatly promoted by an accumulation of heat. This may be effected by means of an extremely strong metallic vessel, called *Papin's digester*, in which the bones and water are enclosed, without any possibility of the steam making its escape. A heat can thus be applied much superior to that of boiling water; and bones, by this means, are completely reduced to a pulp. But the process still consumes too much fuel to be generally adopted among the lower classes.

CAROLINE.

And why should not a manufacture be established for grinding or macerating bones, or at least for reducing them to the state of shavings, when I suppose they would dissolve as readily as hartshorn shavings?

MRS. B.

They could not be collected clean for such a purpose, but they are not lost, as they are used for making hartshorn and sal ammoniac; and such is the superior science and industry of this country, that we now send sal ammoniac to the Levant, though it originally came to us from Egypt.

EMILY.

When jelly is made of isinglass, does it leave no sediment?

MRS. B.

No; nor does it so much require clarifying, as it consists almost entirely of pure gelantine, and any foreign matter that is mixed with it, is thrown off during the boiling in the form of scum.—These are processes which you may see performed in great perfection in the culinary laboratory, by that very able and most useful chemist the cook.

CAROLINE.

To what an immense variety of purposes chemistry is subservient!

EMILY.

It appears, in that respect, to have an advantage over most other arts and sciences; for these, very often, have a tendency to confine the imagination to their own particular object, whilst the pursuit of chemistry is so extensive and diversified, that it inspires a general curiosity, and a desire of enquiring into the nature of every object.

CAROLINE.

I suppose that soup is likewise composed of gelatine; for, when cold, it often assumes the consistence of jelly?

MRS. B.

Not entirely; for though soups generally contain a quantity of gelatine, the most essential ingredient is a mucous or extractive matter, a peculiar animal substance, very soluble in water, which has a strong taste, and is more nourishing than gelatine. The various kinds of portable soup consist of this extractive matter in a dry state, which, in order to be made into soup, requires only to be dissolved in water.

Gelatine, in its solid state, is a semiductile transparent substance, without either taste or smell. —When exposed to heat, in contact with air and water, it first swells, then fuses, and finally burns. You may have seen the first part of this operation performed in the carpenter's glue-pot.

CAROLINE.

But you said that gelatine had no smell, and glue has a very disagreeable one.

MRS. B.

Glue is not pure gelatine; as it is not designed for eating, it is prepared without attending to the state of the ingredients, which are more or less contaminated by particles that have become putrid.

Gelatine may be precipitated from its solution in water by alcohol. — We shall try this experiment with a glass of warm jelly. — You see that the gelatine subsides by the union of the alcohol and the water.

EMILY.

How is it, then, that jelly is flavoured with wine, without producing any precipitation?

MRS. B.

Because the alcohol contained in wine is already combined with water, and other ingredients, and is therefore not at liberty to act upon the jelly as when in its separate state. Gelatine is soluble both in acids and in alkalies; the former, you know, are frequently used to season jellies.

CAROLINE.

Among the combinations of gelatine we must not forget one which you formerly mentioned; that with tannin, to form leather.

MRS. B.

True; but you must observe that leather can be produced only by gelatine in a membranous state; for though pure gelatine and tannin will produce a substance chemically similar to leather, yet the texture of the skin is requisite to make it answer the useful purposes of that substance.

The next animal substance we are to examine is *albumen;* this, although constituting a part of most of the animal compounds, is frequently found insulated in the animal system; the white of egg, for instance, consists almost entirely of albumen; the substance that composes the nerves, the serum, or white part of the blood, and the curds of milk, are little else than albumen variously modified.

In its most simple state, albumen appears in the form of a transparent viscous fluid, possessed of no distinct taste or smell; it coagulates at the low temperature of 165 degrees, and, when once solidified, it will never return to its fluid state.

Sulphuric acid and alcohol are each of them capable of coagulating albumen in the same manner as heat, as I am going to show you.

EMILY.

Exactly so. — Pray, Mrs. B., what kind of action is there between albumen and silver? I have sometimes observed, that if the spoon with which I eat an egg happens to be wetted, it becomes tarnished.

MRS. B.

It is because the white of egg (and, indeed, albumen in general) contains a little sulphur, which, at the temperature of an egg just boiled, will decompose the drop of water that wets the spoon, and produce sulphurated hydrogen gas, which has the property of tarnishing silver.

We may now proceed to *fibrine*. This is an insipid and inodorous substance, having somewhat the appearance of fine white threads adhering together; it is the essential constituent of muscles or flesh, in which it is mixed with and softened by gelatine. It is insoluble both in water and alcohol, but sulphuric acid converts it into a substance very analogous to gelatine.

These are the essential and general ingredients of animal matter; but there are other substances, which, though not peculiar to the animal system, usually enter into its composition, such as oils, acids, salts, &c.

Animal oil is the chief constituent of fat; it is contained in abundance in the cream of milk, whence it is obtained in the form of butter.

EMILY.

Is animal oil the same in its composition as vegetable oils?

MRS. B.

Not the same, but very analogous. The chief difference is that animal oil contains nitrogen, a principle which seldom enters into the composition of vegetable oils, and never in so large a proportion.

There are a few animal acids, that is to say, acids peculiar to animal matter, from which they are almost exclusively obtained.

The animal acids have triple bases of hydrogen, carbon, and nitrogen. Some of them are found native in animal matter; others are produced during its decomposition.

Those that we find ready formed are:

The *bombic acid*, which is obtained from silkworms.

The *formic acid*, from ants.

The *lactic acid*, from the whey of milk.

The *sebacic*, from oil or fat.

Those produced during the decomposition of animal substances by heat, are the *prussic* and *zoonic* acids. This last is produced by the roasting of meat, and gives it a brisk flavour.

CAROLINE.

The class of animal acids is not very extensive?

MRS. B.

No; nor are they, generally speaking, of great importance. The *prussic acid* is, I think, the only one sufficiently interesting to require any further comment. It can be formed by any artificial process, without the presence of any animal matter; and it may likewise be obtained from a variety of vegetables, particularly those of the narcotic kind, such as poppies, laurel, &c. But it is commonly obtained from blood, by strongly heating that substance with caustic potash; the alkali attracts the acid from the blood, and forms with it a *prussiat of potash*. From this state of combination the prussic acid can be obtained pure by means of other substances which have the power of separating it from the alkali.

EMILY.

But if this acid does not exist ready formed in blood, how can the alkali attract it from it?

MRS. B.

It is the triple basis only of this acid that exists in the blood; and this is developed and brought to the state of acid, during the combustion. The acid therefore is first formed, and it afterwards combines with the potash.

EMILY.

Now I comprehend it. But how can the prussic acid be artificially made?

MRS. B.

By passing ammoniacal gas over red-hot charcoal; and hence we learn that the constituents of this acid are hydrogen, nitrogen, and carbon. The two first are derived from the volatile alkali, the last from the combustion of the charcoal.

CAROLINE.

But this does not accord with the system of oxygen being the principle of acidity.

MRS. B.

The colouring matter of prussian blue is called an acid, because it unites with alkalies and metals, and not from any other characteristic properties of acids; perhaps the name is not strictly appropriate. But this circumstance, together with some others of the same kind, has induced several chemists to think that oxygen may not be the exclusive generator of acids. Sir H. Davy, I have already informed you, was led by his experiments on dry acids to suspect that water might be essential to acidity. And it is the opinion of some chemists that acidity may possibly depend rather on the arrangement than on the presence of any particular principles. But we have not yet done with the prussic acid. It has a strong affinity for metallic oxyds, and precipitates the solutions of iron in acids of a blue colour. This is the prussian blue, or prussiat of iron, so much used in the arts, and with which I think you must be acquainted.

EMILY.

Yes, I am; it is much used in painting, both in oil and in water colours; but it is not reckoned a permanent oil-colour.

MRS. B.

That defect arises, I believe, in general, from its being badly prepared, which is the case when the iron is not so fully oxydated as to form a red oxyd. For a solution of green oxyd of iron (in which the metal is more slightly oxydated), makes only a pale green, or even a white precipitate, with prussiat of potash; and this gradually changes to blue by being exposed to the air, as I can immediately show you.

CAROLINE.

It already begins to assume a pale blue colour. But how does the air produce this change?

MRS. B.

By oxydating the iron more perfectly. If we pour some nitrous acid on it, the prussian blue colour will be immediately produced, as the acid will yield its oxygen to the precipitate, and fully saturate it with this principle, as you shall see.

CAROLINE.

It is very curious to see a colour change so instantaneously.

MRS. B.

Hence you perceive that prussian blue cannot be a permanent colour, unless prepared with red oxyd of iron, since by exposure to the atmosphere it gradually darkens, and in a short time is no longer in harmony with the other colours of the painting.

CAROLINE.

But it can never become darker, by exposure to the atmosphere, than the true prussian blue, in which the oxyd is perfectly saturated?

MRS. B.

Certainly not. But in painting, the artist not

reckoning upon partial alterations in his colours, gives his blue tints that particular shade which harmonises with the rest of the picture. If, afterwards, those tints become darker, the harmony of the colouring must necessarily be destroyed.

CAROLINE.

Pray, of what nature is the paint called *carmine?*

MRS. B.

It is an animal colour prepared from *cochineal,* an insect, the infusion of which produces a very beautiful red.

CAROLINE.

Whilst we are on the subject of colours, I should like to learn what *ivory black* is?

MRS. B.

It is a carbonaceous substance obtained by the combustion of ivory. A more common species of black is obtained from the burning of bone.

CAROLINE.

But during the combustion of ivory or bone, the carbon, I should have imagined, must be converted into carbonic acid gas, instead of this black substance?

MRS. B.

In this, as in most combustions, a considerable part of the carbon is simply volatilised by the heat, and again obtained concrete on cooling. This colour, therefore, may be called the soot produced by the burning of ivory or bone.

CONVERSATION XXIV.

ON THE ANIMAL ECONOMY.

MRS. B.

We have now acquired some idea of the various materials that compose the animal system; but if you are curious to know in what manner these substances are formed by the animal organs, from vegetable, as well as from animal substances, it will be necessary to have some previous knowledge of the nature and functions of these organs, without which it is impossible to form any distinct idea of the process of *animalisation* and *nutrition*.

CAROLINE.

I do not exactly understand the meaning of the word animalisation?

MRS. B.

Animalisation is the process by which the food

is *assimilated*, that is to say, converted into animal matter; and nutrition is that by which the food thus assimilated is rendered subservient to the purposes of nourishing and maintaining the animal system.

EMILY.

This, I am sure, must be the most interesting of all the branches of chemistry!

CAROLINE.

So I think; particularly as I expect that we shall hear something of the nature of respiration, and of the circulation of the blood?

MRS. B.

These functions undoubtedly occupy a most important place in the history of the animal economy. — But I must previously give you a very short account of the principal organs by which the various operations of the animal system are performed. These are:

The *Bones*,
Muscles,
Blood vessels,
Lymphatic vessels,
Glands, and
Nerves.

The *bones* are the most solid part of the animal frame, and in a great measure determine its form and dimensions. You recollect, I suppose, what are the ingredients which enter into their composition?

CAROLINE.

Yes; phosphat of lime, cemented by gelatine.

MRS. B.

During the earliest period of animal life, they consist almost entirely of gelatinous membrane having the form of the bones, but of a loose spongy texture, the cells or cavities of which are destined to be filled with phosphat of lime; it is the gradual acquisition of this salt which gives to the bones their subsequent hardness and durability. Infants first receive it from their mother's milk, and afterwards derive it from all animal and from most vegetable food, especially farinaceous substances, such as wheat-flour, which contain it in sensible quantities. A portion of the phosphat, after the bones of the infant have been sufficiently expanded and solidified, is deposited in the teeth, which consist at first only of a gelatinous membrane or case, fitted for the reception of this salt; and which, after acquiring hardness within the gum, gradually protrude from it.

CAROLINE.

How very curious this is; and how ingeniously

nature has first provided for the solidification of such bones as are immediately wanted, and afterwards for the formation of the teeth, which would not only be useless, but detrimental in infancy!

MRS. B.

In quadrupeds the phosphat of lime is deposited likewise in their horns, and in the hair or wool with which they are generally clothed.

In birds it serves also to harden the beaks and the quills of their feathers.

When animals are arrived at a state of maturity, and their bones have acquired a sufficient degree of solidity, the phosphat of lime which is taken with the food is seldom assimilated, excepting when the female nourishes her young; it is then all secreted into the milk, as a provision for the tender bones of the nursling.

EMILY.

So that whatever becomes superfluous to one being, is immediately wanted by another; and the child acquires strength precisely by the species of nourishment which is no longer necessary to the mother. Nature is, indeed, an admirable economist!

CAROLINE.

Pray, Mrs. B., does not the disease in the bones

of children, called the rickets, proceed from a deficiency of phosphat of lime?

MRS. B.

I have heard that this disease may arise from two causes; it is sometimes occasioned by the growth of the muscles being too rapid in proportion to that of the bones. In this case the weight of the flesh is greater than the bones can support, and presses upon them so as to produce a swelling of the joints, which is the great indication of the rickets. The other cause of this disorder is supposed to be an imperfect digestion and assimilation of the food, attended with an excess of acid, which counteracts the formation of phosphat of lime. In both instances, therefore, care should be taken to alter the child's diet, not merely by increasing the quantity of aliment containing phosphat of lime, but also by avoiding all food that is apt to turn acid on the stomach, and to produce indigestion. But the best preservative against complaints of this kind is, no doubt, good nursing: when a child has plenty of air and exercise, the digestion and assimilation will be properly performed, no acid will be produced to interrupt these functions, and the muscles and bones will grow together in just proportions.

CAROLINE.

I have often heard the rickets attributed to bad nursing, but I never could have guessed what connection there was between exercise and the formation of the bones.

MRS. B.

Exercise is generally beneficial to all the animal functions. If man is destined to labour for his subsistence, the bread which he earns is scarcely more essential to his health and preservation than the exertions by which he obtains it. Those whom the gifts of fortune have placed above the necessity of bodily labour are compelled to take exercise in some mode or other, and when they cannot convert it into an amusement, they must submit to it as a task, or their health will soon experience the effects of their indolence.

EMILY.

That will never be my case: for exercise, unless it becomes fatigue, always gives me pleasure; and, so far from being a task, is to me a source of daily enjoyment. I often think what a blessing it is, that exercise, which is so conducive to health, should be so delightful; whilst fatigue, which is rather hurtful, instead of pleasure, occasions painful sensations. So that fatigue, no doubt, was intended to moderate our bodily exertions, as satiety puts a limit to our appetites.

MRS. B.

Certainly. — But let us not deviate too far from our subject. — The bones are connected together by ligaments, which consist of a white thick flexible substance, adhering to their extremities, so far as to secure the joints firmly, though without impeding their motion. And the joints are moreover covered by a solid, smooth, elastic, white substance, called *cartilage*, the use of which is to allow, by its smoothness and elasticity, the bones to slide easily over one another, so that the joints may perform their office without difficulty or detriment.

Over the bones the *muscles* are placed; they consist of bundles of fibres which terminate in a kind of string, or ligament, by which they are fastened to the bones. The muscles are the organs of motion; by their power of dilatation and contraction they put into action the bones, which act as levers, in all the motions of the body, and form the solid support of its various parts. The muscles are of various degrees of strength or consistence in different species of animals. The mammiferous tribe, or those that suckle their young, seem in this respect to occupy an intermediate place between birds and cold-blooded animals, such as reptiles and fishes.

EMILY.

The different degrees of firmness and solidity in the muscles of these several species of animals proceed, I imagine, from the different nature of the food on which they subsist?

MRS. B.

No; that is not supposed to be the case: for the human species, who are of the mammiferous tribe, live on more substantial food than birds, and yet the latter exceed them in muscular strength. We shall hereafter attempt to account for this difference; but let us now proceed in the examination of the animal functions.

The next class of organs is that of the *vessels* of the body, the office of which is to convey the various fluids throughout the frame. These vessels are innumerable. The most considerable of them are those through which the blood circulates, which are of two kinds: the *arteries*, which convey it from the heart to the extremities of the body, and the *veins*, which bring it back into the heart.

Besides these, there are a numerous set of small transparent vessels, destined to absorb and convey different fluids into the blood; they are generally called the *absorbent* or *lymphatic* vessels: but it is to a portion of them only that the function of conveying into the blood the fluid called *lymph* is assigned.

EMILY.

Pray what is the nature of that fluid?

MRS. B.

The nature and use of the lymph have, I believe, never been perfectly ascertained; but it is supposed to consist of matter that has been previously animalised, and which, after answering the purpose for which it was intended, must, in regular rotation, make way for the fresh supplies produced by nourishment. The lymphatic vessels pump up this fluid from every part of the system, and convey it into the veins to be mixed with the blood which runs through them, and which is commonly called venous blood.

CAROLINE.

But does it not again enter into the animal system through that channel?

MRS. B.

Not entirely; for the venous blood does not return into the circulation until it has undergone a peculiar change, in which it throws off whatever is become useless.

Another set of absorbent vessels pump up the *chyle* from the stomach and intestines, and convey it, after many circumvolutions, into the great vein near the heart.

EMILY.

Pray what is chyle?

MRS. B.

It is the substance into which food is converted by digestion.

CAROLINE.

One set of the absorbent vessels, then, is employed in bringing away the old materials that are no longer fit for use; whilst the other set is busy in conveying into the blood the new materials that are to replace them.

EMILY.

What a great variety of ingredients must enter into the composition of the blood?

MRS. B.

You must observe that there is also a great variety of substances to be secreted from it. We may compare the blood to a general receptacle or storehouse for all kinds of commodities, which are afterwards fashioned, arranged, and disposed of as circumstances require.

There is another set of absorbent vessels in females which is destined to secrete milk for the nourishment of the young.

EMILY.

Pray is not milk very analogous in its composition to blood; for, since the nursling derives its nourishment from that source only, it must contain every principle which the animal system requires?

MRS. B.

Very true. Milk is found, by its analysis, to contain the principal materials of animal matter, albumen, oil, and phosphat of lime; so that the suckling has but little trouble to digest and assimilate this nourishment. But we shall examine the composition of milk more fully afterwards.

In many parts of the body numbers of small vessels are collected together in little bundles called *glands*, from a Latin word meaning *acorn*, on account of the resemblance which some of them bear in shape to that fruit. The function of the glands is to *secrete*, or separate certain matters from the blood.

The secretions are not only mechanical, but chemical separations from the blood; for the substances thus formed, though contained in the blood, are not ready combined in that fluid. The secretions are of two kinds, those which form peculiar animal fluids, as bile, tears, saliva, &c.; and those which produce the general materials of the animal system, for the purpose of recruiting and nourishing the several organs of the body; such as albumen, gelatine, and fibrine; the latter may be distinguished by the name of *nutritive secretions.*

CAROLINE.

I am quite astonished to hear that all the secretions should be derived from the blood.

EMILY.

I thought that the bile was produced by the liver?

MRS. B.

So it is; but the liver is nothing more than a very large gland, which secretes the bile from the blood.

The last of the animal organs which we have mentioned are the *nerves;* these are the vehicles of sensation, every other part of the body being, of itself, totally insensible.

CAROLINE.

They must then be spread through every part of the frame, for we are every where susceptible of feeling.

EMILY.

Excepting the nails and the hair.

MRS. B.

And those are almost the only parts in which nerves cannot be discovered. The common source of all the nerves is the brain; thence they descend, some of them through different holes of the skull, but the greatest part through the back bone, and extend themselves by innumerable ramifications throughout the whole body. They spread themselves over the muscles, penetrate the glands, wind round the vascular system, and even pierce into the interior of the bones. It is most probably through them that the communication is carried on between the mind and the other parts of the body; but in what manner they are acted on by the mind, and made to re-act on the body, is still a profound secret. Many hypotheses have been formed on this very obscure subject, but they are all equally improbable, and it would be useless for us to waste our time in conjectures on an enquiry, which, in all probability, is beyond the reach of human capacity.

CAROLINE.

But you have not mentioned those particular nerves that form the senses of hearing, seeing, smelling, and tasting?

MRS. B.

They are considered as being of the same nature as those which are dispersed over every part of the body, and constitute the general sense of feeling. The different sensations which they produce arise from their peculiar situation and connection with the several organs of taste, smell, and hearing.

EMILY.

But these senses appear totally different from that of feeling?

MRS. B.

They are all of them sensations, but variously modified according to the nature of the different organs in which the nerves are situated. For, as we have formerly observed, it is by contact only that the nerves are affected. Thus odoriferous particles must strike upon the nerves of the nose, in order to excite the sense of smelling; in the same manner that taste is produced by the particular substance coming in contact with the nerves of the palate. It is thus also that the sensation of sound is produced by the concussion of the air striking against the auditory nerve; and sight is the effect of the light falling upon the optic nerve. These various senses, therefore, are affected only by the actual contact of particles of matter, in the same manner as that of feeling.

The different organs of the animal body, though easily separated and perfectly distinct, are loosely connected together by a kind of spongy substance, in texture somewhat resembling net-work, called the cellular membrane; and the whole is covered by the skin.

The *skin*, as well as the bark of vegetables, is formed of three coats. The external one is called the *cuticle* or *epidermis*; the second, which is called the *mucous membrane*, is of a thin soft texture, and consists of a mucous substance, which in negroes is black, and is the cause of their skin appearing of that colour.

CAROLINE.

Is then the external skin of negroes white like ours?

MRS. B.

Yes; but as the cuticle is transparent, as well as porous, the blackness of the mucous membrane is visible through it. The extremities of the nerves are spread over this skin, so that the sensation of feeling is transmitted through the cuticle. The internal covering of the muscles, which is properly the skin, is the thickest, the toughest, and most resisting of the whole; it is this membrane which is so essential in the arts, by forming leather when combined with tannin.

The skin which covers the animal body, as well as those membranes that form the coats of the vessels, consists almost exclusively of gelatine; and is capable of being converted into glue, size, or jelly.

The cavities between the muscles and the skin are usually filled with fat, which lodges in the cells of the membranous net before mentioned, and gives to the external form (especially in the human figure) that roundness, smoothness, and softness, so essential to beauty.

EMILY.

And the skin itself is, I think, a very ornamental part of the human frame, both from the fineness of its texture, and the variety and delicacy of its tints.

MRS. B.

This variety and harmonious gradation of colours, proceed, not so much from the skin itself, as from the internal organs which transmit their several colours through it, these being only softened and blended by the colour of the skin, which is uniformly of a yellowish white.

Thus modified, the darkness of the veins appears of a pale blue colour, and the floridness of the arteries is changed to a delicate pink. In the most transparent parts, the skin exhibits the bloom of the rose, whilst where it is more opake its own colour predominates; and at the joints, where the bones are most prominent, their whiteness is often discernible. In a word, every part of the human frame seems to contribute to its external grace; and this not merely by producing a pleasing variety of tints, but by a peculiar kind of beauty which belongs to each individual part. Thus it is to the solidity and arrangement of the bones that the human figure owes the grandeur of its stature, and its firm and dignified deportment. The muscles delineate the form, and stamp it with energy and grace; and the soft substance which is spread over them smooths their ruggedness, and gives to the contours the gentle undulations of the line of beauty. Every organ of sense is a peculiar and separate ornament; and the skin, which polishes the surface, and gives it that charm of colouring so inimitable by art, finally conspires to render the whole the fairest work of the creation.

But now that we have seen in what manner the animal frame is formed, let us observe how it provides for its support, and how the several organs, which form so complete a whole, are nourished and maintained.

This will lead us to a more particular explanation of the internal organs: here we shall not meet with so much apparent beauty, because these parts were not intended by nature to be exhibited to view; but the beauty of design, in the internal organisation of the animal frame, is, if possible, still more remarkable than that of the external parts.

We shall defer this subject till our next interview.

CONVERSATION XXV.

ON ANIMALISATION, NUTRITION, AND RESPIRATION.

MRS. B.

We have now learnt of what materials the animal system is composed, and have formed some idea of the nature of its organisation. In order to complete the subject, it remains for us to examine in what manner it is nourished and supported.

Vegetables, we have observed, obtain their nourishment from various substances, either in their elementary state, or in a very simple state of combination; as carbon, water, and salts, which they pump up from the soil; and carbonic acid and oxygen, which they absorb from the atmosphere.

Animals, on the contrary, feed on substances of the most complicated kind; for they derive their sustenance, some from the animal creation, others from the vegetable kingdom, and some from both.

CAROLINE.

And there is one species of animals, which, not satisfied with enjoying either kind of food in its simple state, has invented the art of combining them together in a thousand ways, and of rendering even the mineral kingdom subservient to its refinements.

EMILY.

Nor is this all; for our delicacies are collected from the various climates of the earth, so that the four quarters of the globe are often obliged to contribute to the preparation of our simplest dishes.

CAROLINE.

But the very complicated substances which constitute the nourishment of animals, do not, I suppose, enter into their system in their actual state of combination?

MRS. B.

So far from it, that they not only undergo a new arrangement of their parts, but a selection is made of such as are most proper for the nourishment of the body, and those only enter into the system, and are animalised.

EMILY.

And by what organs is this process performed?

MRS. B.

Chiefly by the stomach, which is the organ of

digestion, and the prime regulator of the animal frame.

Digestion is the first step towards nutrition. It consists in reducing into one homogeneous mass the various substances that are taken as nourishment; it is performed by first chewing and mixing the solid aliment with the saliva, which reduces it to a soft mass, in which state it is conveyed into the stomach, where it is more completely dissolved by the *gastric juice*.

This fluid (which is secreted into the stomach by appropriate glands) is so powerful a solvent that scarcely any substances will resist its action.

EMILY.

The coats of the stomach, however, cannot be attacked by it, otherwise we should be in danger of having them destroyed when the stomach was empty.

MRS. B.

They are probably not subject to its action; as long, at least, as life continues. But it appears, that when the gastric juice has no foreign substance to act upon, it is capable of occasioning a degree of irritation in the coats of the stomach, which produces the sensation of hunger. The gastric juice, together with the heat and muscular action of the stomach, converts the aliment into an uniform pulpy mass called chyme. This passes into the intestines, where it meets with the bile and some other fluids, by the agency of which, and by the operation of other causes hitherto unknown, the chyme is changed into chyle, a much thinner substance, somewhat resembling milk, which is pumped by immense numbers of small absorbent vessels spread over the internal surface of the intestines. These, after many circumvolutions, gradually meet and unite into large branches, till they at length collect the chyle into one vessel, which pours its contents into the great vein near the heart, by which means the food, thus prepared, enters into the circulation.

CAROLINE.

But I do not yet clearly understand how the blood, thus formed, nourishes the body and supplies all the secretions?

MRS. B.

Before this can be explained to you, you must first allow me to complete the formation of the blood. The chyle may, indeed, be considered as forming the chief ingredient of blood; but this fluid is not perfect until it has passed through the lungs, and undergone (together with the blood that has already circulated) certain necessary changes that are effected by RESPIRATION.

CAROLINE.

I am very glad that you are going to explain the nature of respiration: I have often longed to understand it, for though we talk incessantly of *breathing*, I never knew precisely what purpose it answered.

MRS. B.

It is indeed one of the most interesting processes imaginable; but, in order to understand this function well, it will be necessary to enter into some previous explanations. Tell me, Emily, — what do you understand by respiration?

EMILY.

Respiration, I conceive, consists simply in alternately *inspiring* air into the lungs, and *expiring* it from them.

MRS. B.

Your answer will do very well as a general definition. But, in order to form a tolerably clear notion of the various phenomena of respiration, there are many circumstances to be taken into consideration.

In the first place, there are two things to be distinguished in respiration, the *mechanical* and the *chemical* part of the process.

The mechanism of breathing depends on the alternate expansions and contractions of the chest, in which the lungs are contained. When the chest dilates, the cavity is enlarged, and the air rushes in at the mouth, to fill up the vacuum formed by this dilatation; when it contracts, the cavity is diminished, and the air forced out again.

CAROLINE.

I thought that it was the lungs that contracted and expanded in breathing?

MRS. B.

They do likewise; but their action is only the consequence of that of the chest. The lungs, together with the heart and largest blood vessels, in a manner fill up the cavity of the chest; they could not, therefore, dilate if the chest did not previously expand; and, on the other hand, when the chest contracts, it compresses the lungs and forces the air out of them.

CAROLINE.

The lungs, then, are like bellows, and the chest is the power that works them.

MRS. B.

Precisely so. Here is a curious little figure (Plate XV. Fig. 5., that will assist me in explaining the mechanism of breathing.

CAROLINE.

What a droll figure! a little head fixed upon a glass bell, with a bladder tied over the bottom of it!

MRS. B.

You must observe that there is another bladder within the glass, the neck of which communicates with the mouth of the figure — this represents the lungs contained within the chest; the other bladder, which you see is tied loose, represents a muscular membrane, called the *diaphragm*, which separates the chest from the lower part of the body. By the chest, therefore, I mean that large cavity in the upper part of the body contained within the ribs, the neck, and the diaphragm; this membrane is muscular, and capable of contraction and dilatation. The contraction may be imitated by drawing the bladder tight over the bottom of the receiver, when the air in the bladder, which represents the lungs, will be forced out through the mouth of the figure —

EMILY.

See, Caroline, how it blows the flame of the candle in breathing!

MRS. B.

By letting the bladder loose again, we imitate the dilatation of the diaphragm, and the cavity of the chest being enlarged, the lungs expand, and the air rushes in to fill them.

EMILY.

This figure, I think, gives a very clear idea of the process of breathing.

MRS. B.

It illustrates tolerably well the action of the lungs and diaphragm; but those are not the only powers that are concerned in enlarging or diminishing the cavity of the chest; the ribs are also possessed of a muscular motion for the same purpose; they are alternately drawn in, edgeways, to assist the contraction, and stretched out, like the hoops of a barrel, to contribute to the dilatation of the chest.

EMILY.

I always supposed that the elevation and depression of the ribs were the consequence, not the cause of breathing.

MRS. B.

It is exactly the reverse. The muscular action of the diaphram, together with that of the ribs, are the *causes* of the contraction and expansion of the chest; and the air rushing into, and being expelled from the lungs, are only *consequences* of those actions.

CAROLINE.

I confess that I thought the act of breathing began by opening the mouth for the air to rush in, and that it was the air alone, which, by alternately rushing in and out, occasioned the dilatations and contractions of the lungs and chest.

MRS. B.

Try the experiment of merely opening your mouth; the air will not rush in, till by an interior muscular action you produce a vacuum — yes, just so, your diaphragm is now dilated, and the ribs expanded. But you will not be able to keep them long in that state. Your lungs and chest are already resuming their former state, and expelling the air with which they had just been filled. This mechanism goes on more or less rapidly, but, in general, a person at rest and in health will breathe between fifteen and twenty-five times in a minute.

We may now proceed to the chemical effects of respiration; but, for this purpose, it is necessary that you should previously have some notion of the *circulation* of the blood. Tell me, Caroline, what do you understand by the circulation of the blood?

CAROLINE.

I am delighted that you come to that subject, for it is one that has long excited my curiosity. But I cannot conceive how it is connected with respiration. The idea I have of the circulation is, that the blood runs from the heart through the veins all over the body, and back again to the heart.

MRS. B.

I could hardly have expected a better definition from you; it is, however, not quite correct, for you do not distinguish the *arteries* from the *veins*, which, as we have already observed, are two distinct sets of vessels, each having its own peculiar functions. The arteries convey the blood from the heart to the extremities of the body; and the veins bring it back into the heart.

This sketch will give you an idea of the manner in which some of the principal veins and arteries of the human body branch out of the heart, which may be considered as a common centre to both sets of vessels. The heart is a kind of strong elastic bag, or muscular cavity, which possesses a power of dilating and contracting itself, for the purposes of alternately receiving and expelling the blood, in order to carry on the process of circulation.

EMILY.

Why are the arteries in this drawing painted red, and the veins purple?

MRS. B.

It is to point out the difference of the colour of the blood in these two sets of vessels.

CAROLINE.

But if it is the same blood that flows from the arteries into the veins, how can its colour be changed?

MRS. B.

This change arises from various circumstances. In the first place, during its passage through the arteries, the blood undergoes a considerable alteration, some of its constituent parts being gradually separated from it for the purpose of nourishing the body, and of supplying the various secretions. The consequence of this is, that the florid arterial colour of the blood changes by degrees to a deep purple, which is its constant colour in the veins. On the other hand, the blood is recruited during its return through the veins by the fresh chyle, or imperfect blood, which has been produced by food; and it receives also lymph from the absorbent vessels, as we have before mentioned. In consequence of these several changes, the blood returns to the heart in a state very different from that in which it left it. It is loaded with a greater proportion of hydrogen and carbon, and is no longer fit for the nourishment of the body, or other purposes of circulation.

EMILY.

And in this state does it mix in the heart with the pure florid blood that runs into the arteries?

MRS. B.

No. The heart is divided into two cavities or compartitions, called the *right* and *left ventricles.*

The left ventricle is the receptacle for the pure arterial blood previous to its irculation; whilst the venous, or impure blood, which returns to the heart after having circulated, is received into the right ventricle, previous to its purification, which I shall presently explain.

CAROLINE.

For my part, I always thought that the same blood circulated again and again through the body, without undergoing any change.

MRS. B.

Yet you must have supposed that the blood circulated for some purpose?

CAROLINE.

I knew that it was indispensable to life; but had no idea of its real functions.

MRS. B.

But now that you understand that the blood conveys nourishment to every part of the body, and supplies the various secretions, you must be sensible that it cannot constantly answer these objects without being proportionally renovated and purified.

CAROLINE.

But does not the chyle answer this purpose?

MRS. B.

Only in part. It renovates the nutritive principles of the blood, but does not relieve it from the superabundance of water and carbon with which it is encumbered.

EMILY.

How, then, is this effected?

MRS. B.

By RESPIRATION. This is one of the grand mysteries which modern chemistry has disclosed. When the venous blood enters the right ventricle of the heart, it contracts by its muscular power, and throws the blood through a large vessel into the lungs, which are contiguous, and through which it circulates by millions of small ramifications. Here it comes in contact with the air which we breathe. The action of the air on the blood in the lungs is, indeed, concealed, from our immediate observation; but we are able to form a tolerably accurate judgment of it from the changes which it effects not only in the blood, but also on the air expired.

The air, after passing through the lungs, is found to contain all the nitrogen inspired, but to have lost part of its oxygen, and to have acquired a portion of watery vapour and of carbonic acid gas. Hence it is inferred, that when the air comes in contact with the venous blood in the lungs, the oxygen attracts from it the superabundant quantity of carbon with which it has impregnated itself during the circulation, and converts it into carbonic acid. This gaseous acid, together with the redundant moisture from the lungs *, being then expired, the blood is restored to its former purity, that is, to the state of arterial blood, and is thus again enabled to perform its various functions.

CAROLINE.

This is trùly wonderful! Of all that we have yet learned, I do not recollect any thing that has appeared to me so curious and interesting. I almost believe that I should like to study anatomy now, though I have hitherto had so disgusting an idea of it. Pray, to whom are we indebted for these beautiful discoveries?

MRS. B.

Priestley and Crawford, in this country, and Lavoisier, in France, are the principal inventors of the theory of respiration. Of late years the subject has been farther illustrated and simplified by the accurate experiments of Messrs. Allen and Pepys. But the still more important and more

* The quantity of moisture discharged by the lungs in 24 hours, may be computed at eight or nine ounces.

admirable discovery of the circulation of the blood was made long before by our immortal countryman Harvey.

EMILY.

Indeed I never heard any thing that delighted me so much as this theory of respiration. But I hope, Mrs. B., that you will enter a little more into particulars before you dismiss so interesting a subject. We left the blood in the lungs to undergo the salutary change: but how does it thence spread to all the parts of the body?

MRS. B.

After circulating through the lungs, the blood is collected into four large vessels, by which it is conveyed into the left ventricle of the heart, whence it is propelled to all the different parts of the body by a large artery, which gradually ramifies into millions of small arteries through the whole frame. From the extremities of these little ramifications the blood is transmitted to the veins, which bring it back to the heart and lungs, to go round again and again in the manner we have just described. You see, therefore, that the blood actually undergoes two circulations; the one, through the lungs, by which it is converted into pure arterial blood; the other, or general circulation, by which nourishment is conveyed to every part of the body; and these are both equally indispensable to the support of animal life.

EMILY.

But whence proceeds the carbon with which the blood is impregnated when it comes into the lungs?

MRS. B.

Carbon exists in a greater proportion in blood than in organised animal matter. The blood, therefore, after supplying its various secretions, becomes loaded with an excess of carbon, which is carried off by respiration; and the formation of new chyle from the food affords a constant supply of carbonaceous matter.

CAROLINE.

I wonder what quantity of carbon may be expelled from the blood by respiration in the course of 24 hours?

MRS. B.

It appears by the experiments of Messrs. Allen and Pepys that about 40,000 cubic inches of carbonic acid gas are emitted from the lungs of a healthy person, daily; which is equivalent to *eleven ounces* of solid carbon every 24 hours.

EMILY.

What an immense quantity! And pray how

much of carbonic acid gas do we expel from our lungs at each expiration?

MRS. B.

The quantity of air which we take into our lungs at each inspiration, is about 40 cubic inches, which contain a little less than 10 cubic inches of oxygen; and of those 10 inches, one-eighth is converted into carbonic acid gas on passing once through the lungs*, a change which is sufficient to prevent air which has only been breathed once from suffering a taper to burn in it.

CAROLINE.

Pray, how does the air come in contact with the blood in the lungs?

MRS. B.

I cannot answer this question without entering into an explanation of the nature and structure of the lungs. You recollect that the venous blood, on being expelled from the right ventricle, enters the lungs to go through what we may call the lesser circulation; the large trunk or vessel that conveys it branches out, at its entrance into the lungs, into an infinite number of very fine ramifi-

* The bulk of carbonic acid gas formed by respiration, is exactly the same as that of the oxygen gas which disappears.

cations. The windpipe, which conveys the air from the mouth into the lungs, likewise spreads out into a corresponding number of air vessels, which follow the same course as the blood vessels, forming millions of very minute air-cells. These two sets of vessels are so interwoven as to form a sort of net-work, connected into a kind of spongy mass, in which every particle of blood must necessarily come in contact with a particle of air.

CAROLINE.

But since the blood and the air are contained in different vessels, how can they come into contact?

MRS. B.

They act on each other through the membrane which forms the coats of these vessels; for although this membrane prevents the blood and the air from mixing together in the lungs, yet it is no impediment to their chemical action on each other.

EMILY.

Are the lungs composed entirely of blood vessels and air vessels?

MRS. B.

I believe they are, with the addition only of nerves and of a small quantity of the cellular substance before mentioned, which connects the whole into an uniform mass.

EMILY.

Pray, why are the lungs always spoken of in the plural number? Are there more than one?

MRS. B.

Yes; for though they form but one organ, they really consist of two compartments called *lobes*, which are enclosed in separate membranes or bags, each occupying one side of the chest, and being in close contact with each other, but without communicating together. This is a beautiful provision of nature, in consequence of which, if one of the lobes be wounded, the other performs the whole process of respiration till the first is healed.

The blood, thus completed, by the process of respiration, forms the most complex of all animal compounds, since it contains not only the numerous materials necessary to form the various secretions, as saliva, tears, &c. but likewise all those that are required to nourish the several parts of the body, as the muscles, bones, nerves, glands, &c.

EMILY.

There seems to be a singular analogy between the blood of animals and the sap of vegetables; for each of these fluids contains the several materials destined for the nutrition of the numerous class of bodies to which they respectively belong.

MRS. B.

Nor is the production of these fluids in the animal and vegetable systems entirely different; for the absorbent vessels, which pump up the chyle from the stomach and intestines, may be compared to the absorbents of the roots of plants, which suck up the nourishment from the soil. And the analogy between the sap and the blood may be still further traced, if we follow the latter in the course of its circulation; for, in the living animal, we find every where organs which are possessed of a power to secrete from the blood and appropriate to themselves the ingredients requisite for their support.

CAROLINE.

But whence do these organs derive their respective powers?

MRS. B.

From a peculiar organisation, the secret of which no one has yet been able to unfold. But it must be ultimately by means of the vital principle that both their mechanical and chemical powers are brought into action.

I cannot dismiss the subject of circulation without mentioning *perspiration*, a secretion which is immediately connected with it, and acts a most important part in the animal economy.

CAROLINE.

Is not this secretion likewise made by appropriate glands?

MRS. B.

No; it is performed by the extremities of the arteries, which penetrate through the skin and terminate under the cuticle, through the pores of which the perspiration issues. When this fluid is not secreted in excess, it is *insensible*, because it is dissolved by the air as it exudes from the pores; but when it is secreted faster than it can be dissolved, it becomes *sensible*, as it assumes its liquid state.

EMILY.

This secretion bears a striking resemblance to the transpiration of the sap of plants. They both consist of the most fluid parts, and both exude from the surface by the extremities of the vessels through which they circulate.

MRS. B.

And the analogy does not stop there; for, since it has been ascertained that the sap returns into the roots of the plants, the resemblance between the animal and vegetable circulation is become still more obvious. The latter, however, is far from being complete, since, as we observed before, it consists only in a rising and descending of the sap, whilst in animals the blood actually *circulates* through every part of the system.

We have now, I think, traced the process of nutrition, from the introduction of the food into the stomach to its finally becoming a constituent part of the animal frame. This will, therefore, be a fit period to conclude our present conversation. What further remarks we have to make on the animal economy shall be reserved for our next interview.

CONVERSATION XXVI.

ON ANIMAL HEAT; AND ON VARIOUS ANIMAL PRODUCTS.

EMILY.

SINCE our last interview, I have been thinking much of the theory of respiration; and I cannot help being struck with the resemblance which it appears to bear to the process of combustion. For in respiration, as in most cases of combustion, the air suffers a change, and a portion of its oxygen combines with carbon, producing carbonic acid gas.

MRS. B.

I am much pleased that this idea has occurred to you: these two processes appear so very analogous, that it has been supposed that a kind of combustion actually takes place in the lungs; not of the blood, but of the superfluous carbon which the oxygen attracts from it.

CAROLINE.

A combustion in our lungs! that is a curious idea indeed! But, Mrs. B., how can you call the action of the air on the blood in the lungs combustion, when neither light nor heat are produced by it?

EMILY.

I was going to make the same objection. — Yet I do not conceive how the oxygen can combine with the carbon, and produce carbonic acid, without disengaging heat?

MRS. B.

The fact is, that heat is disengaged.* Whether any light be evolved, I cannot pretend to determine; but that heat is produced in considerable and very sensible quantities is certain, and this is the principal, if not the only source of ANIMAL HEAT.

EMILY.

How wonderful! that the very process which purifies and elaborates the blood, should afford an inexhaustible supply of internal heat?

* It has been calculated that the heat produced by respiration in 12 hours, in the lungs of a healthy person, is such as would melt about 100 pounds of ice.

MRS. B.

This is the theory of animal heat in its original simplicity, such nearly as it was first proposed by Black and Lavoisier. It was equally clear and ingenious; and was at first generally adopted. But it was objected, on second consideration, that if the whole of the animal heat was evolved in the lungs, it would necessarily be much less in the extremities of the body than immediately at its source; which is not found to be the case. This objection, however, which was by no means frivolous, is now satisfactorily removed by the following consideration: — Venous blood has been found by experiment to have *less capacity for heat* than arterial blood; whence it follows that the blood, in gradually passing from the arterial to the venous state, during the circulation, parts with a portion of caloric, by means of which heat is diffused through every part of the body.

EMILY.

More and more admirable!

CAROLINE.

The cause of animal heat was always a perfect mystery to me, and I am delighted with its explanation. — But pray, Mrs. B., can you tell me what is the reason of the increase of heat that takes place in a fever?

EMILY.

Is it no because we then breathe quicker, and therefore more heat is diseng ged in the system?

MRS. B.

That may be one reason: but I should think that the principal cause of the heat experienced in fevers, is, that there is no vent for the caloric which is generated in the body. One of the most considerable secretions is the insensible perspiration; this is constantly carrying off caloric in a latent state; but during the hot stage of a fever, the pores are so contracted, that all perspiration ceases, and the accumulation of caloric in the body occasions those burning sensations which are so painful.

EMILY.

This is, no doubt, the reason why the perspiration that often succeeds the hot stage of a fever affords so much relief. If I had known this theory of animal heat when I had a fever last summer, I think I should have found some amusement in watching the chemical processes that were going on within me.

CAROLINE.

But exercise likewise produces animal heat, and that must be quite in a different manner.

MRS. B.

Not so much so as you think; for the more exercise you take, the more the body is stimulated, and requires recruiting. For this purpose the circulation of the blood is quickened, the breath proportionably accelerated, and consequently a greater quantity of caloric evolved.

CAROLINE.

True; after running very fast, I gasp for breath, my respiration is quick and hard, and it is just then that I begin to feel hot.

EMILY.

It would seem, then, that violent exercise should produce fever.

MRS. B.

Not if the person is in a good state of health; for the additional caloric is then carried off by the perspiration which succeeds.

EMILY.

What admirable resources nature has provided for us! By the production of animal heat she has enabled us to keep up the temperature of our bodies above that of inanimate objects; and whenever this source becomes too abundant, the excess is carried off by perspiration.

MRS. B.

It is by the same law of nature that we are enabled, in all climates, and in all seasons, to preserve our bodies of an equal temperature, or at least very nearly so.

CAROLINE.

You cannot mean to say that our bodies are of the same temperature in summer, and in winter, in England, and in the West-Indies.

MRS. B.

Yes, I do; at least if you speak of the temperature of the blood, and the internal parts of the body; for those parts that are immediately in contact with the atmosphere, such as the hands and face, will occasionally get warmer, or colder, than the internal or more sheltered parts. But if you put the bulb of a thermometer in your mouth, which is the best way of ascertaining the real temperature of your body, you will scarcely perceive any difference in its indication, whatever may be the difference of temperature of the atmosphere.

CAROLINE.

And when I feel overcome by heat, I am really not hotter than when I am shivering with cold?

MRS. B.

When a person in health feels very hot, whether from internal heat, from violent exercise, or from the temperature of the atmosphere, his body is certainly a little warmer than when he feels very cold; but this difference is much smaller than our sensations would make us believe; and the natural standard is soon restored by rest and by perspiration. It is chiefly the external parts that are warmer, and I am sure that you will be surprised to hear that the internal temperature of the body scarcely ever descends below ninety-five or ninety-six degrees, and seldom attains one hundred and four or one hundred and five degrees, even in the most violent fevers.

EMILY.

The greater quantity of caloric, therefore, that we receive from the atmosphere in summer, cannot raise the temperature of our bodies beyond certain limits, as it does that of inanimate bodies, because an excess of caloric is carried off by perspiration.

CAROLINE.

But the temperature of the atmosphere, and consequently that of inanimate bodies, is surely never so high as that of animal heat?

MRS. B.

I beg your pardon. Frequently in the East and West Indies, and sometimes in the southern parts of Europe, the atmosphere is above ninety-eight degrees, which is the common temperature of animal heat. Indeed, even in this country, it occasionally happens that the sun's rays, setting full on an object, elevate its temperature above that point.

In illustration of the power which our bodies have to resist the effects of external heat, Sir Charles Blagden, with some other gentlemen, made several very curious experiments. He remained for some time in an oven heated to a temperature not much inferior to that of boiling water, without suffering any other inconvenience than a profuse perspiration, which he supported by drinking plentifully.

EMILY.

He could scarcely consider the perspiration as an inconvenience, since it saved him from being baked by giving vent to the excess of caloric.

CAROLINE.

I always thought, I confess, that it was from the heat of the perspiration that we suffered in summer.

MRS. B.

You now find that you are quite mistaken. Whenever evaporation takes place, cold, you know, is produced in consequence of a quantity of caloric being carried off in a latent state; this is the case with perspiration, and it is in this way that it affords relief. It is on that account also that we are so apt to *catch cold*, when in a state of profuse perspiration. It is for the same reason that tea is often refreshing in summer, though it appears to heat you at the moment you drink it.

EMILY.

And in winter, on the contrary, tea is pleasant on account of its heat.

MRS. B.

Yes; for we have then rather to guard against a deficiency than an excess of caloric, and you do not find that tea will excite perspiration in winter, unless after dancing, or any other violent exercise.

CAROLINE.

What is the reason that it is dangerous to eat ice after dancing, or to drink any thing cold when one is very hot?

MRS. B.

Because the loss of heat arising from the perspiration, conjointly with the chill occasioned by the cold draught, produce more cold than can be borne with safety, unless you continue to use the same exercise after drinking that you did before; for the heat occasioned by the exercise will counteract the effects of the cold drink, and the danger will be removed. You may, however, contrary to the common notion, consider it as a rule, that cold liquids may, at all times, be drunk with perfect safety, however hot you may feel, provided you are not at the moment in a state of great perspiration, and on condition that you keep yourself in gentle exercise afterwards.

EMILY.

But since we are furnished with such resources against the extremes of heat or cold, I should have thought that all climates would have been equally wholesome.

MRS. B.

That is true, in a certain degree, with regard to those who have been accustomed to them from birth; for we find that the natives of those climates, which we consider as most deleterious, are as healthy as ourselves; and if such climates are unwholesome to those who are habituated to a more moderate temperature, it is because the animal economy does not easily accustom itself to considerable changes.

CAROLINE.

But pray, Mrs. B., if the circulation preserves the body of an uniform temperature, how does it happen that animals are sometimes frozen?

MRS. B.

Because, if more heat be carried off by the atmosphere than the circulation can supply, the cold will finally prevail, the heart will cease to beat, and the animal will be frozen. And, likewise, if the body remained long exposed to a degree of heat, greater than the perspiration could carry off, it would at last lose the power of resisting its destructive influence.

CAROLINE.

Fish, I suppose, have no animal heat, but only partake of the temperature of the water in which they live?

EMILY.

And their coldness, no doubt, proceeds from their not breathing?

MRS. B.

All kinds of fish breathe more or less, though

in a much smaller degree than land animals. Nor are they entirely destitute of animal heat, though, for the same reason, they are much colder than other creatures. They have comparatively but a very small quantity of blood, therefore but very little oxygen is required, and a proportionally small quantity of animal heat is generated.

CAROLINE.

But how can fish breathe under water?

MRS. B.

They breathe by means of the air which is dissolved in the water, and if you put them into water deprived of air by boiling, they are soon suffocated.

If a fish is confined in a vessel of water closed from the air, it soon dies; and any fish put in afterwards would be killed immediately, as all the air had been previouly consumed.

CAROLINE.

Are there any species of animals that breathe more than we do?

MRS. B.

Yes; birds, of all animals, breathe the greatest quantity of air in proportion to their size; and it is to this that they are supposed to owe the peculiar firmness and strength of their muscles, by which they are enabled to support the violent exertion of flying.

This difference between birds and fish, which may be considered as the two extremes of the scale of muscular strength, is well worth observing. Birds residing constantly in the atmosphere, surrounded by oxygen, and respiring it in greater proportions than any other species of animals, are endowed with a superior degree of muscular strength, whilst the muscles of fish, on the contrary, are flaccid and oily; these animals are comparatively feeble in their motions, and their temperature is scarcely above that of the water in which they live. This is, in all probability, owing to their imperfect respiration; the quantity of hydrogen and carbon, that is in consequence accumulated in their bodies, forms the oil which is so strongly characteristic of that species of animals, and which relaxes and softens the small quantity of fibrine which their muscles contain.

CAROLINE.

But, Mrs. B., there are some species of birds that frequent both elements, as, for instance, ducks and other water fowl. Of what nature is the flesh of these?

MRS. B.

Such birds, in general, make but little use of their wings; if they fly, it is but feebly, and only to a short distance. Their flesh, too, partakes of the oily nature, and even in taste sometimes resembles that of fish. This is the case not only with the various kinds of water fowls, but with all other amphibious animals, as the otter, the crocodile, the lizard, &c.

CAROLINE.

And what is the reason that reptiles are so deficient in muscular strength?

MRS. B.

It is because they usually live under ground, and seldom come into the atmosphere. They have imperfect, and sometimes no discernible organs of respiration; they partake therefore of the soft oily nature of fish; indeed, many of them are amphibious, as frogs, toads, and snakes, and very few of them find any difficulty in remaining a length of time under water. Whilst, on the contrary, the insect tribe, that are so strong in proportion to their size, and alert in their motions, partake of the nature of birds, air being their peculiar element, and their organs of respiration being comparatively larger than in other classes of animals.

I have now given you a short account of the principal animal functions. However interesting the subject may appear to you, a fuller investigation of it would, I fear, lead us too far from our object.

EMILY.

Yet I shall not quit it without much regret; for of all the branches of chemistry, it is certainly the most curious and most interesting.

CAROLINE.

But, Mrs. B., I must remind you that you promised to give us some account of the nature of *milk*.

MRS. B.

True. There are several other animal productions that deserve likewise to be mentioned. We shall begin with milk, which is certainly the most important and the most interesting of all the animal secretions.

Milk, like all other animal substances, ultimately yields by analysis oxygen, hydrogen, carbon, and nitrogen. These are combined in it under the forms of albumen, gelatine, oil, and water. But milk contains, besides, a considerable portion of phosphat of lime, the purposes of which I have already pointed out.

CAROLINE.

Yes; it is this salt which serves to nourish the tender bones of the suckling.

MRS. B.

To reduce milk to its elements, would be a very complicated, as well as useless operation; but this fluid, without any chemical assistance, may be decomposed into three parts, *cream*, *curds*, and *whey*. These constituents of milk have but a very slight affinity for each other, and you find accordingly that cream separates from milk by mere standing. It consists chiefly of oil, which being lighter than the other parts of the milk, gradually rises to the surface. It is of this, you know, that butter is made, which is nothing more than oxygenated cream.

CAROLINE.

Butter, then, is somewhat analogous to the waxy substance formed by the oxygenation of vegetable oils.

MRS. B.

Very much so.

EMILY.

But is the cream oxygenated by churning?

MRS. B.

Its oxygenation commences previous to churning, merely by standing exposed to the atmosphere, from which it absorbs oxygen. The process is afterwards completed by churning; the violent motion which this operation occasions brings every particle of cream in contact with the atmosphere, and thus facilitates its oxygenation.

CAROLINE.

But the effect of churning, I have often observed in the dairy, is to separate the cream into two substances, butter and butter-milk.

MRS. B.

That is to say, in proportion as the oily particles of the cream become oxygenated, they separate from the other constituent parts of the cream in the form of butter. So by churning you produce, on the one hand, butter, or oxygenated oil; and, on the other, butter-milk, or cream deprived of oil. But if you make butter by churning new milk instead of cream, the butter-milk will then be exactly similar in its properties to creamed or skimmed milk.

CAROLINE.

Yet butter-milk is very different from common skimmed milk.

MRS. B.

Because you know it is customary, in order to

save time and labour, to make butter from cream alone. In this case, therefore, the butter-milk is deprived of the creamed milk, which contains both the curd and whey. Besides, in consequence of the milk remaining exposed to the atmosphere during the separation of the cream, the latter becomes more or less acid, as well as the butter-milk which it yields in churning.

EMILY.

Why should not the butter be equally acidified by oxygenation?

MRS. B.

Animal oil is not so easily acidified as the other ingredients of milk. Butter, therefore, though usually made of sour cream, is not sour itself, because the oily part of the cream had not been acidified. Butter, however, is susceptible of becoming acid by an excess of oxygen; it is then said to be rancid, and produces the sebacic acid, the same as that which is obtained from fat.

EMILY.

If that be the case, might not rancid butter be sweetened by mixing with it some substance that would take the acid from it?

MRS. B.

This idea has been suggested by Sir H. Davy,

who supposes, that if rancid butter were well washed in an alkaline solution, the alkali would separate the acid from the butter.

CAROLINE.

You said just now that creamed milk consisted of curd and whey. Pray how are these separated?

MRS. B.

They may be separated by standing for a certain length of time exposed to the atmosphere; but this decomposition may be almost instantaneously effected by the chemical agency of a variety of substances. Alkalies, rennet*, and indeed almost all animal substances, decompose milk by combining with the curds.

Acids and spirituous liquors, on the other hand, produce a decomposition by combining with the whey. In order, therefore, to obtain the whey pure, rennet, or alkaline substances, must be used to attract the curds from it.

But if it be wished to obtain the curds pure, the whey must be separated by acids, wine, or other spirituous liquors.

* Rennet is the name given to a watery infusion of the coats of the stomach of a sucking calf. Its remarkable efficacy in promoting coagulation is supposed to depend on the gastric juice with which it is impregnated.

EMILY.

This is a very useful piece of information; for I find white-wine whey, which I sometimes take when I have a cold, extremely heating; now, if the whey were separated by means of an alkali instead of wine, it would not produce that effect.

MRS. B.

Perhaps not. But I would strenuously advise you not to place too much reliance on your slight chemical knowledge in medical matters. I do not know why whey is not separated from curd by rennet, or by an alkali, for the purpose which you mention; but I strongly suspect that there must be some good reason why the preparation by means of wine is generally preferred. I can, however, safely point out to you a method of obtaining whey without either alkali, rennet, or wine; it is by substituting lemon juice, a very small quantity of which will separate it from the curds.

Whey, as an article of diet, is very wholesome, being remarkable light of digestion. But its effect, taken medicinally, is chiefly, I believe, to excite perspiration, by being drunk warm on going to bed.

From whey a substance may be obtained in crystals by evaporation, called *sugar of milk.* This substance is sweet to the taste, and in its composition is so analogous to common sugar, that it is susceptible of undergoing the vinous fermentation.

CAROLINE.

Why then is not wine, or alcohol, made from whey?

MRS. B.

The quantity of sugar contained in milk is so trifling, that it can hardly answer that purpose. I have heard of only one instance of its being used for the production of a spirituous liquor, and this is by the Tartan Arabs; their abundance of horses, as well as their scarcity of fruits, has introduced the fermentation of mares' milk, by which they produce a liquor called *koumiss.* Whey is likewise susceptible of being acidified by combining with oxygen from the atmosphere. It then produces the *lactic acid,* which you may recollect is mentioned amongst the animal acids, as the acid of milk.

Let us now see what are the properties of curds.

EMILY.

I know that they are made into cheese; but I have heard that for that purpose they are separated from the whey by rennet, and yet this you have just told us is not the method of obtaining pure curds?

MRS. B.

Nor are pure curds so well adapted for the formation of cheese. For the nature and flavour of the cheese depend, in a great measure, upon the cream or oily matter which is left in the curds; so that if every particle of cream be removed from the curds, the cheese is scarcely eatable. Rich cheeses, such as cream and Stilton cheeses, derive their excellence from the quantity, as well as the quality, of the cream that enters into their composition.

CAROLINE.

I had no idea that milk was such an interesting compound. In many respects there appears to me to be a very striking analogy between milk and the contents of an egg, both in respect to their nature and their use. They are, each of them, composed of the various substances necessary for the nourishment of the young animal, and equally destined for that purpose.

MRS. B.

There is, however, a very essential difference. The young animal is formed, as well as nourished, by the contents of the egg-shell; whilst milk serves as nutriment to the suckling, only after it is born.

There are several peculiar animal substances which do not enter into the general enumeration of animal compounds, and which, however, deserve to be mentioned.

Spermaceti is of this class; it is a kind of oily substance obtained from the head of the whale, which, however, must undergo a certain preparation before it is in a fit state to be made into candles. It is not much more combustible than tallow, but it is pleasanter to burn, as it is less fusible and less greasy.

Ambergris is another peculiar substance derived from a species of whale. It is, however, seldom obtained from the animal itself, but is generally found floating on the surface of the sea.

Wax, you know, is a concrete oil, the peculiar product of the bee, part of the constituents of which may probably be derived from flowers, but so prepared by the organs of the bee, and so mixed with its own substance, as to be decidedly an animal product. Bees' wax is naturally of a yellow colour, but it is bleached by long exposure to the atmosphere, or may be instantaneously whitened by the oxy-muriatic acid. The combustion of wax is far more perfect than that of tallow, and consequently produces a greater quantity of light and heat.

Lac is a substance very similar to wax in the manner of its formation; it is the product of an insect, which collects its ingredients from flowers, apparently for the purpose of protecting its eggs from injury. It is formed into cells, fabricated with as much skill as those of the honey-comb, but differently arranged. The principal use of lac is in the manufacture of sealing-wax, and in making varnishes and lacquers.

Musk, *civet*, and *castor*, are other particular productions, from different species of quadrupeds. The two first are very powerful perfumes; the latter has a nauseous smell and taste, and is only used medicinally.

CAROLINE.

Is it from this substance that castor oil is obtained?

MRS. B.

No. Far from it, for castor oil is a vegetable oil, expressed from the seeds of a particular plant; and has not the least resemblance to the medicinal substance obtained from the castor.

Silk is a peculiar secretion of the silk-worm, with which it builds its nest or cocoon. This insect was originally brought to Europe from China. Silk, in its chemical nature, is very similar to the hair and wool of animals; whilst in the insect it is a fluid, which is coagulated, apparently by uniting with oxygen, as soon as it comes in contact with the air. The moth of the silk-worm ejects a liquor which appears to contain a particular acid, called *bombic*, the properties of which are but very little known.

EMILY.

Before we conclude the subject of the animal economy, shall we not learn by what steps dead animals return to their elementary state?

MRS. B.

Animal matter, although the most complicated of all natural substances, returns to its elementary state by one single spontaneous process, the *putrid fermentation.* By this, the albumen, fibrine, &c. are slowly reduced to the state of oxygen, hydrogen, nitrogen, and carbon; and thus the circle of changes through which these principles have passed is finally completed. They first quitted their elementary form, or their combination with unorganised matter, to enter into the vegetable system. Hence they were transmitted to the animal kingdom; and from this they return again to their primitive simplicity, soon to re-enter the sphere of organised existence.

When all the circumstances necessary to produce fermentation do not take place, animal, like vegetable matter, is liable to a partial or imperfect decomposition, which converts it into a combustible substance very like spermaceti. I dare say that Caroline, who is so fond of analogies, will consider this as a kind of animal bitumen.

CAROLINE.

And why should I not, since the processes which produce these substances are so similar?

MRS. B.

There is, however, one considerable difference; the state of bitumen seems permanent, whilst that of animal substances, thus imperfectly decomposed, is only transient; and unless precautions be taken to preserve them in that state, a total dissolution infallibly ensues. This circumstance, of the occasional conversion of animal matter into a kind of spermaceti, is of late discovery. A manufacture has in consequence been established near Bristol, in which, by exposing the carcases of horses and other animals for a length of time under water, the muscular parts are converted into this spermaceti-like substance. The bones afterwards undergo a different process to produce hartshorn, or, more properly, ammonia, and phosphorus; and the skin is prepared for leather.

Thus art contrives to enlarge the sphere of useful purposes, for which the elements were intended by nature; and the productions of the several kingdoms are frequently arrested in their course, and variously modified, by human skill, which compels them to contribute, under new forms, to the necessities or luxuries of man.

But all that we enjoy, whether produced by the

spontaneous operations of nature, or the ingenious efforts of art, proceed alike from the goodness of Providence. — To God alone man owes the admirable faculties which enable him to improve and modify the productions of nature, no less than those productions themselves. In contemplating the works of the creation, or studying the inventions of art; let us, therefore, never forget the Divine Source from which they proceed; and thus every acquisition of knowledge will prove a lesson of piety and virtue.

INDEX.

A

Absorbent vessels, ii. 304
Absorption of caloric, i. 59. 66
Acetic acid, ii. 75. 197
Acetous fermentation, ii. 232
——— acid, ii. 193. 232
Acidulous gaseous mineral waters, ii. 129
——— salts, ii. 200
Acids, i. 262. ii. 69
Aeriform, i. 36
Affinity, i. 19. ii. 1
Agate, ii. 51
Agriculture, ii. 252
Air, i. 182. ii. 262
Albumen, ii. 277. 288
Alburnum, ii. 267
Alchemists, i. 4
Alcohol, or spirit of wine, ii. 215. 222
Alembic, i. 258
Alkalies, ii. 19
Alkaline earths, ii. 50. 58
Alloys, i. 344
Alum, or sulphat of alumine, ii. 55. 95
Alumine, ii. 54
Alumium, i. 13
Amalgam, i. 347
Ambergris, ii. 358
Amethyst, ii. 58
Amianthus, ii. 66
Ammonia, or volatile alkali, 363. i. 20. 35
Ammoniacal gas, ii. 36
Ammonium, i. 13
Analysis, i. 287
——— of vegetables, ii. 165
Animals, ii. 276
Animal acids, ii. 75. 290
——— colours, ii. 292
——— heat, ii. 337
——— oil, ii. 178. 283
Animalization, ii. 276. 297. 315
Antidotes, ii. 41. 87
Antimony. i. 14
Aqua fortis, ii. 105
—— regia, i. 340. ii. 144
Arrack, ii. 220
Argand's Lamp, i. 208
Arsenic, i. 14. 340. 348
Arteries, ii. 304. 323
Arterial blood, ii. 305. 326. 338
Asphaltum, ii. 240
Assafœtida, ii. 188
Assimilation, ii. 298
Astringent principle, ii. 198
Atmosphere, i. 90. 181. ii. 262
Atmospherical air, i. 182
Attraction of aggregation, or cohesion, i. 16. ii. 2

Attraction of composition, i. 16. ii. 1
Azot, or nitrogen, i. 182. ii. 100
Azotic gas, i. 182

B

Balsams, ii. 165. 188
Balloons, i. 245
Bark, ii. 193. 265
Barytes, ii. 44. 58. 61
Bases of acids, i. 263. ii. 69
——— gases, i. 183
——— salts, ii. 5
Beer, ii. 212. 220
Benzoic acid, ii. 74. 197
Bile, ii. 308
Birds, ii. 347
Bismuth, i. 14
Bitumens, ii. 239
Black lead, or plumbago, i. 304
Bleaching, i. 32. ii. 89. 146.
Blow-pipe, i. 324. ii. 226
Blood, ii. 306. 317
Blood-vessels, ii. 298
Boiling water, i. 93
Bombic acid, ii. 75. 290
Bones, ii. 298, 299
Boracic acid, i. 365. ii. 131
Boracium, i. 13. ii. 132
Borat of soda, ii. 133
Brandy, ii. 218
Brass, i. 344
Bread, ii. 233
Bricks, ii. 56
Brittle-metals, i. 14
Bronze, i. 344
Butter, ii. 351
Butter-milk, ii. 352

C

Calcareous earths, ii. 65
——— stones, ii. 123
Calcium, i. 13
Caloric, i. 12. 33
———, absorption of, i. 66
———, conductors of, i. 70
———, combined, i. 122
———, expansive power of i. 35
———, equilibrium of, i. 50
———, reflexion of, i. 54. 67
———, radiation of, i. 52. 61
———, solvent power of, i. 96. 102
———, capacity for, i. 124
Calorimeter, i. 156
Calx, i. 183
Camphor, ii. 165. 185
Camphoric acid, ii. 74. 197
Caoutchouc, ii. 165. 189
Carbonats, ii. 25. 129
Carbonat of ammonia, ii. 41
——— lead, i. 320
——— lime, ii. 59. 130
——— magnesia, ii. 67
——— potash, ii. 25
Carbonated hydrogen gas, i. 302
Carbon, i. 282. ii. 329
Carbonic acid, i. 290. 359. ii. 327
Carburet of iron, i. 304. 342
Carmine, ii. 295
Cartilage, ii. 303
Castor, ii. 359
Cellular membrane, ii. 311
Caustics, i. 349
Chalk, ii. 62. 123
Charcoal, i. 282
Cheese, ii. 356
Chemical attraction, i. 15. ii. 9
Chemistry, i. 3
Chest, ii. 318
China, ii. 54
Chlorine, i. 214
Chrome, i. 14. 340
Chyle, ii. 305. 317
Chyme, ii. 316

Citric acid, ii. 74. 197
Circulation of the blood, ii. 322
Civet, ii. 359
Clay, i. 48. ii. 55
Coke, ii. 241
Coal, ii. 240. 252
Cobalt, i. 14
Cochineal, ii. 295
Cold, i. 50. 58
—— from evaporation, 102. 113. 150
Colours of metallic oxyds, i. 319
Columbium, i. 14. 340. 348
Combined caloric, i. 122
Combustion, i. 190
——, volatile products of, i. 207
——, fixed products of, i. 207
——, of alcohol, ii. 225
——, of ammoniacal gas, ii. 42
——, of boracium, ii. 133
——, by oxymuriatic acid or chlorine, ii. 142
——, of carbon, i. 289
——, of coals, i. 207. 297
——, of charcoal by nitric acid, ii. 102
——, of candles, i. 236. 309. ii. 179
——, of diamonds, i. 292
——, of ether, ii. 230
——, of hydrogen, i. 229. 235. 247
——, of iron, i. 200. 322
——, of metals, i. 321
——, of oils, i. 208. ii. 178. 309
——, of oil of turpentine by nitrous acid, ii. 6
——, of phosphorus, i. 272
——, of sulphur, i. 261
Combustion of potassium, i. 358. ii. 132. 138, 139
Compound bodies, i. 9. ii. 14
—— or neutral salts i. 333. ii. 4
Conductors of heat, i. 71
——, solids, i. 73
——, fluids, i. 78
——, Count Rumford's theory, i. 79
Constituent parts, i. 9
Copper, i. 14. 331
Copal, ii. 187. 224
Cortical layers, ii. 265. 267
Cotyledons, or lobes, ii. 256
Cream, ii. 351
Cream of tartar, or tartrit of potash, ii. 200. 222
Cryophorus, i. 154
Crystallisation, i. 338. ii. 47
Cucurbit, i. 258
Culinary heat, 88
Curd, ii. 351. 354
Cuticle, or epidermis, ii. 310

D

Decomposition, i. 8. 20
—— of atmospherical air, i. 181. 209
—— of water by the Voltaic battery, i. 220
—— of salts by the Voltaic battery, ii. 14
—— of water by metals, i. 225. 334
—— by carbon, i. 301
—— of vegetables, ii. 202
—— of potash, i. 356
—— of soda, i. 56
—— of ammonnia, 363. ii. 37

Decomposition of the boracic acid, ii. 132
———— of the fluoric acid, ii. 136
———— of the muriatic acid, ii. 139
Deflagration, ii. 118
Definite proportions, ii. 13
Deliquescence, ii. 95
Detonation, i. 219. ii. 116
Dew, i. 105
Diamond, i. 285
Diaphragm, ii. 320
Digestion, ii. 316
Dissolution of metals, i. 165. 316. 333
Distillation, i. 259. ii. 218
———— of red wine, ii. 218
Divellent forces, ii. 12
Division, i. 7
Drying oils, ii. 181
Dying, ii. 191

E

Earths, ii. 44
Earthen-ware, ii. 53. 57
Effervescence, i. 298
Efflorescence, ii. 94
Elastic fluids, i. 37
Electricity, i. 12. 25. 160. 220. ii. 139
Electric machine, i. 169
Elective attractions, ii. 9
Elementary bodies, i. 8. 12
Elixirs, tinctures, or quintessences, ii. 225
Enamel, ii. 57
Epidermis of vegetables, ii. 269
———— of animals, ii. 310
Epsom salts, ii. 63. 95
Equilibrium of caloric, i. 50
Essences, i. 307. ii. 183. 224
Essential, or volatile oils, i. 307. ii. 183
Ether, i. 111. ii. 229
Evaporation, i. 103
Evergreens, ii. 274
Eudiometer, i. 276
Expansion of caloric, i. 36
Extractive colouring matter, ii. 165. 190

F

Falling stones, i. 319
Fat, i. 306. ii. 311
Feathers, ii. 300
Fecula, ii. 176
Fermentation, ii. 205
Fibrine, ii. 277. 289
Fire, i. 7. 27
Fish, ii. 346
Fixed air, or carbonic acid, i. 290. ii. 125
—— alkalies, ii. 20
—— oils, i. 307. ii. 165. 177
—— products of combustion, i. 207
Flame, i. 237
Flint, ii. 30 51
Flower or blossom, ii. 271
Fluoric acid, ii. 54. 134
Fluorium, or Fluorine, i. 12. ii. 136
Formic acid, ii. 290
Fossil wood, ii. 242
Francincense, ii. 187
Free or radiant caloric, or heat of temperature, i. 33
Freezing mixtures, i. 142
———— by evaporation, 104. 150, &c.
Frost, i. 94
Fruit, ii. 271
Fuller's earth, ii. 55
Furnace, i. 304

G

Galls, ii. 199
Gallat of iron, ii. 98
Gallic acid, ii. 74. 197, 198
Galvanism, i. 163
Gas, i. 182
Gas-lights, i. 240
Gaseous oxyd of carbon, i. 296
——— nitrogen, ii. 111
Gastric juice, ii. 316
Gelatine, or jelly, ii. 277. 280
Germination, ii. 256
Gin, ii. 221
Glands, ii. 298. 307
Glass, ii. 30
Glauber's salts, or sulphat of soda, ii. 92
Glazing, ii. 57
Glucium, i. 13
Glue, 281. 287
Gluten, ii. 165. 177
Gold, i. 14. 323
Gum, ii. 170
—— arabic, ii. 170
—— elastic, or caoutchouc, ii. 189
—— resins, ii. 165. 188
Gunpowder, ii. 116
Gypsum, or plaister of Paris, or sulphat of lime, ii. 95

H

Hair, ii. 300
Harrogate water, i. 268. 341
Hartshorn, ii. 35. 39. 281. 285
Heart, ii. 323
—— wood, ii. 268
Heat, i. 26. 33
—— of capacity, i. 127. 135
—— of temperature, i. 33
Honey, ii. 175
Horns, ii. 282. 300
Hydro-carbonat, i. 241. 303
Hydrogen, i. 214
——— gas, i. 215

I

Jasper, ii. 51
Ice, i. 138
Jelly, ii. 281
Jet, ii. 240
Ignes fatui, i. 277
Ignition, i. 119
Imponderable agents, i. 12
Inflammable air, i. 215
Ink, ii. 98. 199
Insects, ii. 349
Integrant parts, i. 9
Iridium, i. 14
Iron, i. 14. 319. 328
Isinglass, ii. 194. 285
Ivory black, ii. 295
Iodine, i. 214. ii. 157

K

Kali, ii. 34
Koumiss, ii. 356

L

Lac, ii. 358
Lactic acid, ii. 75. 290. 356
Lakes, colours, ii. 190
Latent heat, i. 133
Lavender water, ii. 184. 224
Lead, i. 14. 318. 330
Leather, ii. 193. 287
Leaves, ii. 260
Life, ii. 159. 168
Ligaments, ii. 303
Light, i. 12. 26. ii. 261
Lightning, i. 248
Lime, ii. 59
—— water, ii. 61

Limestone, ii. 60
Linseed oil, ii. 178
Liqueurs, ii. 224
Liver, ii. 308
Lobes, ii. 256. 332
Lunar caustic ,or nitrat of silver, i. 350. ii. 119
Lungs, ii. 319. 330
Lymph, ii. 304
Lymphatic vessels, ii. 304

M

Magnesia, ii. 44. 66
Magnium, i. 13
Malic acid, ii. 74. 197
Malt, ii. 211
Malleable metals, i. 14
Manganese, i. 14. 317
Manna, ii. 176
Manure, ii. 247
Marble, ii. 123
Marine acid, or muriatic acid, ii. 136
Mastic, ii. 187. 224
Materials of animals, ii. 277
——— of vegetables, ii. 165
Mercury, i. 14. 346
———, new mode of freezing, i. 155. 347
Metallic acids, i. 340
——— oxyds, i. 316
Metals, i. 12. 314
Meteorie stones, i. 342
Mica, ii. 66
Milk, ii. 299. 306. 350
Minerals, i. 315. ii. 44. 158
Mineral waters, i. 296. ii. 129
——— acids, ii. 73
Miner's lamp, i. 249
Mixture, i. 99
Molybdena, i. 14. 340
Mordant, ii. 165. 192
Mortar, ii. 53. 65
Mucilage, ii. 170
Mucous acid, ii. 74. 171. 197
——— membrane, ii. 311
Muriatic acid, or marine acid, ii. 136
Muriats, ii. 151
Muriat of ammonia, ii. 35. 152
——— lime, i. 100
——— soda, or common salt, ii. 136. 151
——— potash, ii. 138
Muriatium, i. 13
Muscles of animals, ii. 298. 303
Musk, ii. 359
Myrrh, ii. 188

N.

Naphtha, i. 357. ii. 240
Negative electricity, i. 25. 161. 185
Nerves, ii. 279. 298. 308
Neutral, or compound salts, i. 333. ii. 4. 22. 69
Nickel, i. 13. 343
Nitre, or nitrat of potash, or salt-petre, ii. 32. 104. 116
Nitric acid, ii. 100
Nitrogen, or azot, i. 181. ii. 100
——— gas, i. 182. 211
Nitro-muriatic acid, or aqua regia, ii. 144
Nitrous acid gas, ii. 101. 106
——— air, or nitrit oxyd gas, ii. 107
Nitrats, ii. 116
Nitrat of copper, ii. 5
——— ammonia, ii. 113. 118
——— potash, or nitre, or salt-petre, ii. 32. 104. 116
——— silver, or lunar caustic, ii. 19
Nomenclature of acids, i. 264. ii. 69
——————— compound salts, ii. 4. 22

Nomenclature of other binary compounds, i. 278
Nut-galls, ii. 98. 199
Nut-oil, ii. 178
Nutrition, ii. 297

O

Ochres, i. 320
Oils, i. 285. ii. 306
Oil of amber, ii. 241
—— vitriol, or sulphuric acid, ii. 80
Olive oil, ii. 178
Ores, i. 315
Organized bodies, ii. 159
Organs of animals, ii. 290. 310
—— vegetables, ii. 159. 265. 271
Osmium, i. 14. 348
Oxalic acid, ii. 74. 197
Oxyds, i. 198
Oxyd of manganese, i. 117. 317
—— iron, i. 204. 319
—— lead, i. 319
—— sulphur, ii. 91
Oxydation, or oxygenation, i. 196
Oxygen, i. 11. 181. 201. 211
—— gas, or vital air, i. 182. 201
Oxy-muriatic acid, ii. 140
Oxy-muriats, ii. 153
Oxy-muriat of potash, ii. 153

P

Palladium, i. 13. 348
Papin's digester, i. 120. ii. 284
Parenchyma, ii. 256. 266
Particles, i. 16
Pearlash, ii. 24
Peat, ii. 242
Peculiar juice of plants, ii. 268
Perfect metals, i. 14. 32
Perfumes, i. 308. ii. 183
Perspiration, ii. 333. 339
Petrification, ii. 237
Pewter, i. 344
Pharmacy, i. 14
Phosphat of lime, ii. 99. 299
Phosphorated hydrogen ga 277
Phosphorescence, i. 29
Phosphoric acid, i. 273. ii. 99
Phosphorous acid, i. 274. ii. 99
Phosphorus, i. 270
Phosphoret of lime, i. 278. 341
—— sulphur, i. 279. 341
Pitch, ii. 187
Plaster, ii. 65
Platina, i. 14. 323
Plating, i. 345
Plumbago, or black lead, i. 304
Plumula, ii. 257
Porcelain, ii. 56
Positive electricity, i. 25. 161. 185
Potassium, i. 13. 357. ii. 15
Pottery, ii. 56
Potash, i. 356. ii. 22
Precipitate, i. 22
Pressure of the atmosphere, i. 112. 116
Printer's ink, ii. 144
Prussiat of iron, or prussian blue, ii. 291
—— potash, ii. 291
Prussic acid, ii. 75. 290
Putrid fermentation, ii. 235. 360
Pyrites, i. 341. ii. 97
Pyrometer, i. 38. 42

Q

Quick lime, ii. 59
Quiescent forces, ii. 12

R

Radiation of caloric, i. 52
———, Prevost's theory, i. 52
———, Pictet's explanations, i. 54
———, Leslie's illustrations, i. 61
Radicals, ii. 5. 69
Radicle; or root, ii. 257
Rain, i. 104
Rancidity, ii. 182
Rectification, ii. 223
Reflexion of caloric, i. 54. 64
Reptiles, ii. 349
Resins, ii. 165, 186. 266
Respiration, ii. 317 .326
Reviving of metals, i. 327
Rhodium, i. 14. 348
Roasting metals, i. 316
Rock crystal, ii. 61
Ruby, ii. 53
Rum, ii. 219
Rust, i. 318. 328

S

Saccharine fermentation, ii. 208
Sal ammoniac, or muriat of ammonia, ii. 35
—- polychrest, or sulphat of potash, ii. 91
—- volatile, or carbonat of ammonia, ii. 41
Salifiable bases, ii. 5
Salifying principles, ii. 5
Saltpetre, or nitre, or nitrat of potash, ii. 32. 104. 116
Salt, ii. 91
Sand, ii. 30. 51
Sandstone, ii. 51
Sap of plants, ii. 165. 260. 262. 270. 272
Sapphire, ii. 58
Saturation, i. 101.
Sapphire, ii. 58
Saturation, i. 101
Seas, temperature of, i. 33.
Sebacic acid, ii. 75. 182. 290. 353
Secretions, ii. 307
Seeds of plants, ii. 210. 271
Seltzer water, i. 289. ii. 63. 129
Senses, ii. 310
Silex, or silica, ii. 30. 51
Silicium, i. 13.
Silk, ii. 359
Silver, i. 321
Simple bodies, i. 10. 12
Size, ii. 281
Skin, ii. 279. 310. 193
Slakeing of lime, i. 147. ii. 56
Slate, ii. 51. 66
Smelting metals, i. 316
Smoke, i. 208
Soap, ii. 24
Soda, i. 363. ii. 33
—— water, i. 299
Sodium, i. 13. 363
Soils, i. 42. ii. 245
Soldering, i. 345
Solubility, ii. 92
Solution, i. 96
——- by the air, i. 102
——- of potash, ii. 28
Specific heat, i. 126
Spermaceti, ii. 358
Spirits, ii. 313
Steam, i. 140. 182
Steel, i. 305
Stomach, ii. 315
Stones, ii. 46
Stucco, ii. 65
Strontites, ii. 44. 68
Strontium, i. 13
Suberic acid, ii. 74. 197
Sublimation, i. 257
Succin, or yellow amber, ii. 241
Succinic acid, ii. 74. 197. 241
Sugar, ii. 165. 174. 208
——- of milk, ii. 355

Sulphats, ii. 5. 91
Super oxygenated sulphuric acid, ii. 70.
Sulphat of alumine, or alum, ii. 54. 95
——— barytes, ii. 58
——— iron, ii. 96
——— lime, or gypsum of plaster of Paris, ii. 95
——— magnesia, or Epsom salt, ii. 67. 95
——— potash, or sal polychrest, ii. 91
——— soda, or Glauber's salts, ii. 92
Sulphur, i. 256
——— flowers of, 257
Sulphurated hydrogen gas, i. 165. 268
Sulphurets, i. 341
Sulphurous acid, i. 254. ii. 88
Sulphuric acid, i. 74. ii. 265
Sympathetic ink, i. 354
Synthesis, i. 287

T

Tan, ii. 192
Tannin, ii.165. 192
Tar, ii. 187
Tartarous acid, ii. 74. 197
Tartrit of potash, ii. 222
Teeth, ii. 300
Tellurium, i. 14
Temperature, i. 33
Thaw, i. 158
Thermometers, i. 40
———, Fahrenheit's, i. 42
———, Reaumur's, i. 42
———, Centrigade, i. 43
———, air, i. 44
———, differentiial, i. 46
Thunder, i. 248
Tin, i. 14. 344
Titanium, i. 14. 348
Turf, ii. 242
Turpentine, ii. 187
Transpiration of plants, ii. 260
Tungsten, i. 14. 340

V

Vapour, i. 36. 49. 93. 182
Vaporisation, i. 103
Varnishes, ii. 187
Vegetables, ii. 158
Vegetable acid, i. 310. ii. 74. 197
——— colours, ii. 190
——— heat, ii. 272
——— oils, ii. 177
Veins, ii. 304. 323.
Venous blood, ii. 305. 326. 338
Ventricles, ii. 324
Verdigris, i. 352
Vessels, ii. 304
Vinegar, ii. 232
Vinous fermentation, ii. 212
Vital air, or oxygen gas, i. 182
Vitriol, or sulphat of iron, ii. 81
Volatile oils, i. 307. ii. 165. 183. 224. 269
——— products of combustion, i. 207
——— alkali, i. 363. ii. 20. 35
Voltaic battery, i. 164. 220. 356 ii. 15

U

Uranium, i. 14

W

Water, i. 215. ii. 262

Water, decomposition of, by electricity, i. 200. 225
——, condensation of, i. 32
—— of the sea, i. 86]
——, boiling, i. 93
——, solution by, i. 96
—— of crystallisation, i. 339
Wax, i. 309. ii. 180. 358
Whey, ii. 351
Wine, ii. 212
Wood, ii. 267
Woody fibre, ii. 156. 196. 267
Wool, ii. 300

Y

Yeast, ii. 234.
Yttria, ii. 44.
Yttrium, i. 13.

Z

Zinc, ii. 14. 344
Zicornia, ii. 44
Zicornium, i. 13.
Zoonic acid, ii. 75. 290

END.

Printed by A. Strahan,
Printers-Street, London.

Printed in the United States
By Bookmasters